Springer Series in Computational Physics

Ole G. Mouritsen

Computer Studies of Phase Transitions and Critical Phenomena

With 79 Figures

Springer-Verlag
Berlin Heidelberg New York Tokyo 1984

Dr. Ole G. Mouritsen

Aarhus University, Department of Chemistry,
Physical Chemistry Division, Langelandgade 140,
DK-8000 Aarhus C, Denmark

Editors

H. Cabannes

Mécanique Théoretique
Université Pierre et Marie Curie
Tour 66. 4, Place Jussieu
F-75005 Paris, France

M. Holt

College of Engineering and
Mechanical Engineering
University of California
Berkeley, CA 94720, USA

H. B. Keller

Applied Mathematics 101-50
Firestone Laboratory
California Institute of Technology
Pasadena, CA 91125, USA

J. Killeen

Lawrence Livermore Laboratory
P.O. Box 808
Livermore, CA 94551, USA

S. A. Orszag

Department of Mathematics
Massachusetts Institute of Technology
Cambridge, MA 02139, USA

V. V. Rusanov

Keldysh Institute of Applied Mathematics
4 Miusskaya pl.
SU-125047 Moscow, USSR

ISBN-13: 978-3-642-69711-1   e-ISBN-13: 978-3-642-69709-8
DOI: 10.1007/978-3-642-69709-8

Offset printing: Beltz Offsetdruck, 6944 Hemsbach/Bergstr. Bookbinding: J. Schäffer OHG, 6718 Grünstadt.
2153/3130-5 4 3 2 1 0

*The most successful computer
experiment is the one
which makes itself superfluous*

Preface

This book is based on research carried out by the author in close collaboration with a number of colleagues. In particular, I wish to thank Per Bak, A. John Berlinsky, Hans C. Fogedby, Barry Frank, S. J. Knak Jensen, David Mukamel, David Pink, and Martin Zuckermann for fruitful and extremely stimulating cooperation. It is a pleasure for me to note that active interaction with most of these colleagues is still continuing.

The work has been performed at several different institutions, notably the Department of Chemistry, Aarhus University, Denmark, and the Department of Physics, University of British Columbia, Canada. I wish to thank the Department of Chemistry at Aarhus University for providing me with splendid research facilities over the years. From May 1980 to August 1981, I visited the Department of Physics at the University of British Columbia and I would like to express my sincere gratitude to members of the department for providing me with excellent working conditions. My special thanks are due to Professor Myer Bloom who introduced me to the field of phase transitions in biological membranes and in whose biomembrane group I found an extremely stimulating scientific atmosphere happily married with a most agreeable social climate.

During the last two years when a major part of this work was carried out, I was supported by A/S De Danske Spritfabrikker through their Jubilæumslegat of 1981. Their support is gratefully acknowledged.

The manuscript for this book has been typeset by means of the *TEX* system* as implemented at *RECAU* (The Regional Computing Center of Aarhus University). I am greatly indebted to Marianne Nørnberg for her enthusiasm in piloting this project in mathematical typesetting. I wish to thank her for diligence and authority in typing the manuscript into the computer – and for her patient struggles to get it out again. Steen Larsen and Benedict Løfstedt of *RECAU* are sincerely thanked for indispensable guidance in using *TEX,* for

---

* TEX is a typesetting system intended especially for the creation of mathematical books. The system was invented by prof. Donald E. Knuth, the University of Stanford, California. The TEX implementation used in producing this monograph is based on the TEX May 1980 version from Stanford, adapted to RECAU's CDC Cyber environment by Erik Bertelsen, RECAU. The typesetting device used is a Compugraphic MCS 8600 phototypesetter installed at RECAU.

designing the formats, and for valuable aid in expediting the typesetting. I appreciate the technical assistance of Arne Lindahl in preparing the illustrations.

Finally, my thanks are owed to Kirsten for her constant encouragement and for bearing with me. I thank her for a thorough criticism of my technicalese and other linguistic oddities. I wish to dedicate this book to my son Jonas who did his best to make it impossible.

Aarhus, Denmark, June 1984   *Ole G. Mouritsen*

Table of Contents

1. Introduction

The present survey consists of two parts. The first part outlines the objectives and methods in computer studies of phase transitions and critical phenomena. The fundamental problems inherent in a theoretical description of cooperatively behaving many-body systems will be formulated, and it will be described how computer studies can be useful in solving these problems. General principles will be emphasized more than technical details.

In the second part, a number of applications of computer techniques to specific physical problems will be presented. Rather than attempting to give a full review of this rapidly expanding and outbranching disipline, I have chosen to focus on a small number of examples which are selected to represent the very broad range of possible applications. Consequently, without losing the broad perspective of the approach, this choice offers the advantage of treating each example in some detail and thereby, hopefully, it brings the reader in a position to evaluate the usefulness and potentials of a computer-based approach to cooperative phenomena.

Computer studies of phase transitions and critical phenomena have previously been reviewed extensively by Binder (1976). The wide applicability of computer simulations to many branches of statistical physics is succinctly demonstrated in *Monte Carlo Methods in Statistical Physics I* (1979) and *II* (1984) edited by Kurt Binder. The present survey complements these reviews by giving an expanded treatment of a selection of physical problems of current interest.

What do I mean by »computer studies« and what justifies the use of the term at all? In some way, it seems irrelevant to label a discipline by the name of the calculating device which has been employed in the study. However, it is important to keep in mind that the introduction of fast large-scale electronic computers to statistical physics has had a tremendous and in some sense unsurpassed impact on the whole field. In this work, by computer studies I shall imply theoretical approaches to calculating properties of systems undergoing phase transitions, in which the use of a fast computer is *essential* to simulate numerically or calculate exactly solutions to certain statistical mechanical equations. These solutions could not normally have been obtained without spending astronomically large amounts of time and manpower. Although the computer is essential to the field, I do hope to demonstrate that one need neither be a computer scientist nor know very much about computers to carry out useful computer studies of interesting and non-trivial physical problems. As in all other branches of physics, physical imagination and intuition are the scientist's most important prerequisites.

The specific computer studies chosen for the second part are mainly selected among my own contributions to the field. They represent and illustrate the three main classes of applications of *numerical* computer simulation to cooperative phenomena:

(i)   pure theoretical model calculations in statistical mechanics

(ii)   assessment of the validity of assumptions and predictions of modern theories of phase transitions and critical phenomena

(iii) direct numerical experiments on model systems.

Some of the applications also demonstrate the use of computers to evaluate *exact* configurational lattice statistics. This type of statistics allows construction of perturbation expansions from which information on critical behavior may be extracted by asymptotic analysis.

The main purpose of presenting this selection of applications is to demonstrate how diverse phenomena in a great variety of physical, chemical, and biological systems can be approached succesfully by powerful computer methods. The systems I deal with range from electronic and nuclear magnetic systems, adsorbed molecular monolayers, and alloys to wet lipid bilayers and biological membranes. The kind of phenomena which will be discussed range from magnetic structure and symmetry, general thermodynamic properties, phase transitions and critical behavior to kinetics of growth, lipid bilayer melting, and thermodynamics of lipid-protein interactions in biological membranes. Special attention will be paid to biological membranes since those systems have not before been included in surveys of computer studies. The main point which will be made is to emphasize the importance of applying physical principles to biological research.

2. Computer Methods in the Study of Phase Transitions and Critical Phenomena

Among the most spectacular and remarkable macroscopic events in nature are transformations between the various states of matter. Such transformations, or phase transitions, have fascinated scientists for centuries. A systematic knowledge about phase transitions found an early application in chemical technology and in material sciences such as metallurgy. In recent years, the field of phase transitions has expanded enormously. From embracing transformations between the classical states of matter, i.e. the solid, liquid, and gaseous phases, the field now includes transitions to a variety of phases characterized by such diverse properties as super-conductivity, superfluidity, magnetic ordering, surface structures, ferroelectricity, cosmological quark confinement, chaos, topological ordering, helix-coiling of proteins, and fluidity of biological membranes.

Phase transitions are characterized by abrupt changes, discontinuities, and strong fluctuations. It has been known for a long time that such singular behavior is a consequence of a cooperative phenomenon and thus intimately related to the interactions between the microscopic constituents of matter. The number of constituents in a macroscopic system is typically of the order of 10^{23} and each constituent may carry several degrees of freedom. It is therefore obvious that a theoretical description of phase transitions, which takes the microscopic world as its starting point, is going to be very difficult. The theoretical basis, which claims that such a description is in principle possible, is provided by *statistical mechanics*. Statistical mechanics may be thought of as a set of rules which gives the precise mathematical relationship between the microscopic and macroscopic (thermodynamic) descriptions of a physical system. The usefulness of statistical mechanics rests on the fact that the fundamental physical laws governing the interactions between the microscopic constituents of physical systems are believed to be known. The rules are formulated mathematically in terms of equations involving multi-dimensional integrals. Only in very special cases can these integrals be evaluated analytically exact. Approximation schemes have to be introduced.

It is at this point that the modern computer enters as an indispensable tool to make progress. By using a fast computer, the integrals of statistical mechanics may be evaluated with an accuracy which is only limited by the computer power available. In *experimental* sciences it has been commonly accepted for many years that computers enter at basically every stage of an experiment. However, it is only very recently that the intellectual respectability of applying a computer as an essential part of a *theoretical* development has been widely acknowledged. In this work, we shall address two theoretical approaches to the study of phase transitions, approaches which are conceptually very different but which both base themselves on the use of a computer.

The first approach is of a purely numerical nature. It uses the computer to *simulate* directly the behavior of a physical system taking as its starting point the fundamental equations of statistical mechanics. It is like an experiment: it is built on as little bias as possible. Only the fundamental physical laws governing the interaction between the microscopic constituents are invoked. Contrary to a real experiment, the simulation is carried out on a well-defined system and there is full control over every experimental parameter. However, the simulation shares with the real experiment an important potential, namely that it allows for new discoveries which could not trivially be inferred from the basic physical laws of interaction. The outcome of a simulation may be thought of both as new experimental data and as a theoretical result which can be used to assess the validity of basic assumptions and predictions of analytical theories. Thus, computer simulation interpolates between theory and experiment. By this unique ability, computer simulation may serve to illustrate and illuminate subtle and, unfortunately, not commonly recognized basic conceptual relations in scientific reasoning. There exist two major types of computer simulation techniques which have proved exceedingly successful in the study of phase transitions, namely molecular dynamics methods and Monte Carlo methods. In a molecular dynamics simulation, the deterministic time-evolution of the system is calculated by a numerical integration of the Newtonian equations of motion. In a Monte Carlo simulation, certain stochastic elements are introduced which facilitate the evaluation of the statistical mechanical equations. The principles of applying Monte Carlo simulation techniques to study phase transitions will be described in Sec. 2.2.

The second approach is of an analytical nature. It uses the computer simply as a means of calculating a deterministic statistical mechanical equation to get some numbers out. Full consideration of the physics involved in the problem has been given in advance and any simplifying ideas are embodied in the initial equation. No qualitatively new discoveries will emerge from the calculation itself. The resulting numbers may be compared with the results of theories and with experiments to evaluate the validity of the considerations and the simplifying ideas underlying the initial equation. Perturbation expansions may be constructed by such an approach. The idea is to expand analytically the initial intractable integral in terms of some suitable expansion parameter (e.g. the temperature) and some simpler integrals which can be evaluated exactly on the computer. The high-temperature series expansions described in Sec. 2.3 are obtained by this type of computer-based approach to phase transitions.

When used in combination with each other and in conjunction with modern theories, the two computer-based approaches described above represent the most powerful theoretical tools for studying phase transitions.

Before we proceed with a detailed description of the two computer-based approaches, (Secs. 2.2 and 2.3), we have interposed some general remarks (Sec. 2.1.1) on the theoretical treatment of phase transitions and critical phenomena. These remarks serve to lodge computer studies within the overall framework of theoretical approaches. Furthermore, in Sec. 2.1.2 is given a set of basic statistical mechanical relations. Order parameters and critical exponents will be defined and the concepts of scaling and universality are introduced. This section is completely standard and should make a convenient reference for the general reader.

2.1 Statistical Mechanics and Phase Transitions

2.1.1 Modern theories of phase transitions and critical phenomena

The singularities associated with a critical phenomenon are caused by fluctuations. These fluctuations not only persist on microscopic length scales but cover all length scales up to macroscopic wave lengths. None of these scales can be neglected. To be specific, let us as an example consider the critical phenomenon in water which takes place at 218 atm and 374 °C. Water and steam can be distinguished by their different density. At the critical point, however, the fluctuations in the density cover all length scales and the two phases cannot be distinguished. The observable consequence of this phenomenon is that steam bubbles of all sizes will occur. This is in sharp contrast to what is observed at the ordinary boiling point at 1 atm and 100 °C at which water is transformed into steam by a first-order phase transition. No dramatic fluctuations accompany this transition and the steam bubbles are confined to small sizes due to a finite surface tension.

There exists a number of other physical phenomena which are also influenced by fluctuations of all wave lengths. In his Nobel Lecture, Wilson (1983) for example mentions turbulent fluid flow, internal structure of elementary particles, and the interaction between electrons in a metal and magnetic impurities. All these problems are known to be extremely difficult and they have presented major challenges to theorists for many years.

The modern theoretical approaches to phase transitions and critical phenomena may be divided into two basic groups. The first group consists of direct solution methods applied to models. The physical properties of the models are calculated by means of statistical mechanics. The calculations are performed analytically or numerically, and the solutions of the models may be exact or approximate. The second group consists of methods which exclusively exploit symmetries of the models under consideration. These approaches do not provide solutions of the models in a literal sense but merely attempt to examine how the parameters of the models change under certain scale transformations. From the nature of these changes, the critical properties are derived.

From the first group stem the few very outstanding exact solutions of non-trivial models (e.g. the two-dimensional Ising model (Onsager 1944) and the eight-vertex model (Baxter 1971)). This group also includes various effective-field theoretical approaches (e.g. the celebrated mean-field theory (Smart 1966)) which provides the simplest possible approximate solutions of statistical mechanical models exhibiting phase transitions. Of major importance in the first group are the various approximate calculation schemes normally used in many-body theory, e.g. perturbation expansions such as high- and low-temperature series expansions (Domb and Green 1974). Finally, the first group of approaches contains numerical simulation methods like molecular dynamics (Rahman 1964, Erpenbeck and Wood 1977, Kushick and Berne 1977) and Monte Carlo methods (Metropolis et al. 1953, Binder 1976, 1979, 1984) which during the last decade have developed into a very effective and important tool in the tool-box of the theoretical physicist. Until the introduction of the renormalization group to critical phenomena by Wilson (1971,

1983), analysis of series expansions and numerical simulation were virtually the only methods which yielded reliable information on phase transitions and critical phenomena of non-trivial and exactly unsolved models.

The renormalization group approach (Wilson and Kogut 1974, Domb and Green 1976, Wilson 1983) is the most important element in the second group. In one decade, the renormalization group has developed into perhaps the most important concept in the theory of critical phenomena. Though the renormalization group theory still lacks a firm mathematical foundation (Griffiths 1981), it has had enormous success. In particular, it has provided a coherent picture of universality of critical phenomena (cf. Sec. 2.1.2). A very recent development has been the combination of real-space renormalization group schemes and Monte Carlo techniques as suggested by Ma (1976) and Swendsen (1979) (see Swendsen 1982 for a recent review). Such combined approaches, especially when implemented on special-purpose computers (Hirsch and Scalapino 1983), appear to be extremely promising.

2.1.2   Statistical mechanics, order parameters, fluctuations, critical exponents, scaling, and universality

A microstate, or configuration, of a system is described by a set of mechanical variables, $\bar{\Omega}$. $\bar{\Omega}$ contains the values of all possible degrees of freedom for each particle of the system, e.g. spatial position, velocity, and magnetic moment. The *phase space*, $\{\bar{\Omega}\}$, is the space spanned by all possible microstates of a system. The properties of the system is governed by a Hamiltonian function, $H(\bar{\Omega})$, defined on the mechanical variables. This *Hamiltonian*, which is usually said to define the system, couples the various mechanical variables included in $\bar{\Omega}$. By statistical mechanics, a probability is associated with each microstate. This probability may be expressed in terms of a canonical density function

$$\rho(\bar{\Omega}) = \frac{e^{-H(\bar{\Omega})/k_B T}}{Z}, \tag{2.1.1}$$

where Z is a normalization factor (the partition function)

$$Z = \int_{\{\bar{\Omega}\}} e^{-H(\bar{\Omega})/k_B T} \, d\bar{\Omega}. \tag{2.1.2}$$

T is the absolute temperature and k_B is Boltzmann's constant. Given the probability distribution of the microstates, the thermodynamic value of a measurable physical quantity, $f(\bar{\Omega})$, is obtained in the canonical ensemble as

$$<f> = Z^{-1} \int_{\{\bar{\Omega}\}} f(\bar{\Omega})\rho(\bar{\Omega}) \, d\bar{\Omega}. \tag{2.1.3}$$

Equation (2.1.3) constitutes the formal connection between the microscopic and macroscopic physical worlds.

A particularly interesting situation arises when the partition function Z exhibits singularities. Such singularities may physically be associated with phase transitions

(Yang and Lee 1952). Using the definition of the free energy

$$F = -k_B T \ln Z, \tag{2.1.4}$$

we shall classify phase transitions as *continuous* or *first-order transitions* according to whether first derivatives of F are continuous or not at the singular point, i.e. at the phase transition. This classification, which has been suggested by Fisher (1960), is more convenient than the classic Ehrenfest classification.

A phase transition is characterized by a spontaneously broken symmetry. The phase with the broken symmetry has a symmetry which is lower than that of the Hamiltonian. Symmetry-breaking is conveniently described in terms of an *order parameter*, Φ. We shall here restrict ourselves to systems with phases which can be described by long-range order parameters with a global symmetry.[*] For a given problem, there is no single way of choosing Φ. Its magnitude measures the degree of long-range order in some way. It has the symmetry of the ordered phase and it may have several components. To the order parameter can be associated a thermodynamic conjugate field, h_Φ, which couples directly to Φ. The corresponding term in the free energy is $-h_\Phi \Phi$. Consequently, the order parameter is defined by $\Phi = -(\partial F/\partial h_\Phi)_T$. According to the Fisher classification of phase transitions, we then see that Φ is discontinuous at a first-order transition and goes to zero continuously at a continuous transition (critical point). In experiments as well as in computer simulations, the behavior of the order parameter at the phase transition is usually a key to determine the nature of the phase transition. Broken symmetry is a special case of *broken ergodicity* (Palmer 1982). For most of the systems with ordered phases which we shall discuss in this work, the time scale of observation is so short that the ergodicity is effectively broken. This imposes some very important restrictions on the use of the canonical description to calculate ensemble averages. We shall discuss this delicate point in detail in Sec. 2.2.5.

The susceptibility of the order parameter is an important quantity which is expected to be influenced by critical fluctuations. The isothermal ordering susceptibility, χ_Φ, is defined by

$$\chi_\Phi = (\partial \Phi/\partial h_\Phi)_T. \tag{2.1.5}$$

For a classical system, the fluctuation theorem always relates fluctuations in the order parameter to the corresponding susceptibility. Introducing the order parameter operator M, we therefore have

$$\chi_\Phi = (k_B T)^{-1}(<M^2> - \Phi^2), \tag{2.1.6}$$

where $\Phi = <M>$. This is an important result for computer simulations because it implies that response functions may be evaluated directly from the thermal fluctuations and a perturbing ordering field need not to be introduced. Another useful version of the fluctuation theorem relates the fluctuations in internal energy to the specific heat

---

[*] This may be contrasted to systems with a local gauge symmetry.

$$C_h = (\partial E/\partial T)_h = (k_B T^2)^{-1}(<H^2> - <H>^2). \qquad (2.1.7)$$

Widom (1965) was the first to advance the *scaling hypothesis* for static critical phenomena.[*] According to this hypothesis, the singular part, $\tilde{F}$, of the free energy is a generalized homogeneous function

$$\tilde{F}(\lambda^a t, \lambda^b h_\Phi) = \lambda \tilde{F}(t, h_\Phi), \qquad (2.1.8)$$

where $t \equiv (T - T_c)/T_c$ measures the relative distance from the critical temperature, T_c. Taking the derivative of $\tilde{F}$ with respect to h_Φ in order to obtain the corresponding singular behavior of the order parameter, we find

$$\Phi(t, h = 0) \sim (-t)^\beta \qquad (2.1.9)$$

with the critical exponent $\beta = (1 - b)/a$. Similarly, power-law singularities may be derived for other quantities, e.g.

$$\begin{aligned} C_h(t) &\sim t^{-\alpha}, \quad T > T_c \\ &\sim (-t)^{-\alpha'}, \quad T < T_c \end{aligned} \qquad (2.1.10)$$

$$\begin{aligned} \chi_\Phi(t) &\sim t^{-\gamma}, \quad T > T_c \\ &\sim (-t)^{-\gamma'}, \quad T < T_c. \end{aligned} \qquad (2.1.11)$$

Scaling implies that not all critical exponents are independent (Stanley 1971). Indeed, only two of them can be chosen independently (e.g. a and b in Eq. (2.1.8)). This two-factor scale invariance is usually expressed in terms of socalled scaling relations, e.g.

$$\begin{aligned} \alpha' + 2\beta + \gamma' &= 2 \\ \alpha &= \alpha' \\ \gamma &= \gamma'. \end{aligned} \qquad (2.1.12)$$

It is now known that corrections-to-scaling are important away from the limit $t \to 0$. For example, the order parameter has the following form

$$\Phi(t) \simeq B(-t)^\beta \left(1 + \sum_{i=1} a_i |t|^{\theta_i}\right), \qquad (2.1.13)$$

where the θ_i are correction-to-scaling exponents (Wegner 1972, Aharony and Fisher 1983).

A fundamental observation has been, experimentally as well as theoretically, that the critical exponent values are rather insensitive to details within the system displaying the critical phenomenon. This observation is embodied in the *universality hypothesis* which states that continuous phase transitions can be classified

---

[*] We shall here restrict ourselves to discuss scaling of static critical phenomena. Dynamic scaling is discussed by Halperin and Hohenberg (1977). Scaling for first-order transitions is only little developed (Fisher and Berker 1982).

in a few universality classes, each class giving rise to a certain set of exponents. These classes are determined by a few very fundamental properties of the systems, such as spatial dimension (d), range of interaction, and the symmetry and dimensionality (n) of the order parameter. The physical idea underlying the universality hypothesis is that at a critical point, all details of the microscopic interactions are washed out by the long wave-length fluctuations. It has been the main merit of the renormalization group theory to provide a sound mathematical foundation to the concept of universality.

2.2 Numerical Simulation Techniques

2.2.1 Monte Carlo methods

Numerical simulation of approximate solutions to statistical problems is a fairly old game well-known to the experimental mathematician (see e.g. Meyer 1956, Hammersley and Handscomb 1967). Since numerical simulation methods are concerned with experiments on random numbers, these methods are often called *Monte Carlo* methods. To gain the full power of numerical simulation techniques, a large amount of random numbers has to be generated and processed. Therefore, it is only with the appearance of fast modern computers that the use of Monte Carlo methods has really gained impetus.

Monte Carlo methods are used to solve numerically mathematical problems which are too complex to allow an exact analytical treatment. The problems approached by Monte Carlo methods are conveniently divided into two classes consisting of probabilistic and deterministic problems, respectively. In solving a probabilistic problem, one tries to simulate directly the random process inherent in the problem. Classical examples are simulation of neutron diffusion in reactors and simulation of the random fluctuations in a telephone network. Solving a deterministic problem by a Monte Carlo calculation requires a transformation of the deterministic problem into another problem of a stochastic nature. The original problem need not itself have anything to do with random processes. The only requirement is that the original problem and the transformed one have solutions which differ by a controlled amount. Examples of deterministic problems which have been solved by Monte Carlo methods are differential equations in electromagnetism and multi-dimensional integrals in many-body theory.

Monte Carlo methods can be used with various degrees of sophistication. The most convenient and efficient way of implementing the methods depends to a large extent on the precise problem under consideration. Traditionally, different Monte Carlo methods are distinguished by the sampling techniques they employ. A sampling technique is in turn characterized by the bias imposed on the sampling scheme. One type of sampling scheme is *importance-sampling* which, briefly, is a scheme to collect statistical information according to its importance for a particular problem.

Here, we shall be concerned with the type of Monte Carlo methods which allow a numerical evaluation of the multi-dimensional integrals (cf. Eq. (2.1.3)) which arise in statistical mechanical treatments of interacting many-body systems. A certain Monte Carlo importance-sampling method first proposed by Metropolis et al. (1953) has proved particularly successful in statistical mechanics. The method and its realizations are amazingly simple but nevertheless extremely powerful. The use of the method has had a tremendous impact on the whole field of statistical mechanics. In the following, we shall refer to this method as *the* Monte Carlo method. (Other sampling schemes, e.g. that of Alexandrowicz (Meirovitch 1982), are described by Binder (1979).) In Chaps. 3 - 5, it will be demonstrated that by this very simple method, extremely valuable information can be obtained on a great variety of physical, chemical, and biological problems — information which often is not accessible by any other means. In particular, it will be shown that the difficult problems related to phase transitions may be approached by this method.

As implied by Eq. (2.1.1), we shall use the canonical description of statistical mechanics. Furthermore, we shall restrict ourselves to classical systems with particles. Only lattice models with no translational degrees of freedom will be considered. The sphere of application for the Monte Carlo method is not limited to these cases, however. In fact, the method may be adapted to quantum systems (see e.g. Kolb 1983 and references therein, Schmidt and Kalos 1984) and to systems with translational degrees of freedom (Levesque et al. 1984). Also, other statistical ensembles may be simulated, e.g. the microcanonical and grand canonical ensembles. A particularly elegant illustration of microcanonical Monte Carlo simulation and of how an interpolation to canonical sampling may be set up has recently been described by Creutz (1983).

The type of simulations described and employed in the present work will be termed *conventional* Monte Carlo simulation. This is in order to distinguish it from Monte Carlo renormalization group simulation. The latter kind of simulation is a combination of conventional Monte Carlo simulation and a real-space renormalization group transformation (Swendsen 1982).

In Secs. 2.2.2 - 2.2.11 will be given an account of conventional Monte Carlo techniques as applied to cooperative phenomena. This account is not intended to be a complete description of the various techniques.[*] Rather, the description has been tailored to serve as the background necessary to appreciate the technical aspects of the various applications presented in Chaps. 3 - 5.

2.2.2   A Monte Carlo importance-sampling method

Let us consider a physical system for which the Hamiltonian is known and for which the energy of a given microstate, $E_i = H(\overline{\Omega}_i)$, is easy to calculate. (The latter will be the case for most non-quantum systems.) The properties of the system in thermodynamic equilibrium are determined by Eq. (2.1.3) which constitutes the fundamental mathematical object for the following discussion. For simplicity,

---

[*] See e.g. Binder (1979, 1984) for a more complete and systematic description.

let us assume that the phase space is discrete, $\{\overline{\Omega}_i\}_{i=1}$. (A continuous phase space can be approximated by a discrete one by a suitable division into cells.) Equation (2.1.3) then reads

$$<f> = Z^{-1} \sum_i f(\overline{\Omega}_i)\rho(\overline{\Omega}_i), \qquad (2.2.1)$$

which is still intractable analytically. The trick is now to introduce stochastic elements into the calculation. The simplest possible procedure would involve an unbiased choice of a uniformly random set of points in phase space (random sampling, crude Monte Carlo). However, this procedure would be highly inefficient since the Boltzmann weights vary many orders of magnitude in the neighborhood of phase transitions. Therefore, we shall rather impose a certain bias and choose a finite set of points, $\{\overline{\Omega}'_i\}_{i=1}^M$, in which each state, $\overline{\Omega}'_i$, occurs with a frequency proportional to its Boltzmann probability, Eq. (2.1.1). Thus, microstates are sampled according to their importance (importance-sampling).* Using this particular set, we obtain the (microcanonical) estimate

$$<f> \simeq M^{-1} \sum_{i=1}^{M} f(\overline{\Omega}'_i). \qquad (2.2.2)$$

The particular set, or ensemble, $\{\overline{\Omega}'_i\}_{i=1}^M$, is generated by setting up a random walk in phase space. To this random walk we associate a *Markov chain*, which in this case is discrete and can be parameterized by a discrete time parameter, t, the Markov time. The sequence of states, $\overline{\Omega}'_1, \overline{\Omega}'_2, \ldots, \overline{\Omega}'_M$, is a realization of the Markov chain which is determined by an initial configuration $\overline{\Omega}'_1$ and a stochastic matrix $\overline{\overline{P}}$ with elements p_{ij}, $1 \leq i,j \leq M$, subject to the conditions

$$\forall i,j : p_{ij} \geq 0, \quad \forall i : \sum_{j=1}^{M} p_{ij} = 1. \qquad (2.2.3)$$

p_{ij} denotes the conditional probability for the one-step transition $\overline{\Omega}'_i \rightarrow \overline{\Omega}'_j$ in the chain. We shall here only be concerned with homogeneous Markov chains defined by time-independent transition probabilities. To replace the configurational average in Eq. (2.2.1) by the »time«-average in Eq. (2.2.2), we must construct $\overline{\overline{P}}$ so as to make the limit distribution of the chain, π_j, equal to $\rho(\overline{\Omega}_j)$. If we define the n-step transition probabilities $p_{ij}^{(n)}$ as

$$\forall i,j : p_{ij}^{(n)} = \sum_{k=1}^{M} p_{ik}^{(n-1)} p_{kj}; \quad p_{ij}^{(1)} = p_{ij}, \qquad (2.2.4)$$

---

* This biased selection scheme is from the point of view of variance-reduction not the optimal one. In fact, the optimal scheme is in computational terms extremely impractical and is therefore never used in statistical mechanics since it requires a self-consistent determination of every average $<f>$ to be calculated (Fosdick 1963).

this requirement can be formulated as

$$\forall j: \lim_{n \to \infty} p_{ij}^{(n)} = \pi_j > 0. \tag{2.2.5}$$

According to the theory for homogeneous Markov chains (see e.g. Feller 1950), this limit exists if the chain is irreducible (i.e. all elements are in the same ergodic class). The limit distribution is then independent of j (and therefore independent of the initial state, $t = 1$) and determined uniquely by the normalization and steady-state conditions

$$\sum_{j=1}^{M} \pi_j = 1 \tag{2.2.6}$$

$$\forall j: \pi_j = \sum_{i=1}^{M} \pi_i p_{ij}. \tag{2.2.7}$$

Usually, Eq. (2.2.7) is fulfilled by imposing the stronger condition of microscopic reversibility

$$\forall j, k: \pi_j p_{jk} = \pi_k p_{kj}. \tag{2.2.8}$$

The above conditions imposed on the stochastic matrix $\overline{\overline{P}}$ imply $2M$ linear equations. Thus, we have a considerable freedom left in choosing the M^2 matrix elements. This is computationally extremely fortunate.

Metropolis et al. (1953) have suggested a particular scheme for choosing $\overline{\overline{P}}$, which in a very simple and ingenious way avoids the problem that Z is unknown. The scheme involves an arbitrary symmetric stochastic matrix $\overline{\overline{P}}^*$ (with elements p_{ij}^*) with an associated irreducible ergodic Markov chain. In terms of $\overline{\overline{P}}^*$, the matrix $\overline{\overline{P}}$ is defined

$$p_{ij} = p_{ij}^* \quad , \quad \pi_j/\pi_i \geq 1$$
$$p_{ij} = p_{ij}^* \pi_j/\pi_i, \quad \pi_j/\pi_i < 1 \tag{2.2.9}$$

$$p_{ii} = 1 - \sum_{k(\neq i)}^{M} p_{ik}. \tag{2.2.10}$$

It is easy to prove that these transition probabilities satisfy Eqs. (2.2.3) and (2.2.8). The important point to note is that only ratios of the π's enter the formalism. Obviously, several variations on this theme immediately suggest themselves. It should also be noted that the various options in particular realizations of the above Monte Carlo method are now collected in $\overline{\overline{P}}^*$.

The presentation of the Monte Carlo method in terms of a Markov chain with a time-parameter makes it easy to visualize that the process of generating a chain of microstates may be given a dynamical interpretation. In fact, the Markov process described above is governed by the master equation

$$\frac{d\pi_i(t)}{dt} = -\sum_j p_{ij}\pi_i(t) + \sum_j p_{ji}\pi_j(t). \tag{2.2.11}$$

In thermal equilibrium, $d\pi_i(t)/dt = 0$, and $\lim_{t\to\infty} \pi_i(t) = \rho(\bar{\Omega}_i)$. It is important to note, however, that this dynamics is not the true physical dynamics of the system since the correct equations of motion have not been invoked. Nevertheless, in many cases the time-evolution given by the Markov chain closely resembles the kinetics of the approach to thermodynamic equilibrium (see e.g. Binder and Kalos 1979). The characteristic time scale of the dynamics is determined by $\bar{\bar{P}}^*$, i.e, $\tau^{-1} \sim p_{ij}^*$. The optional character of $\bar{\bar{P}}^*$ permits various constraints to be imposed on the dynamics. For example, $\bar{\bar{P}}^*$ can be chosen so as to make a certain quantity a constant of motion.

Although the method devised by Metropolis et al. is now more than thirty years old, it still thrives and constitutes the main core of most Monte Carlo work in statistical mechanics.

2.2.3   A realization of a Monte Carlo method

We now proceed to describe possible realizations of the Monte Carlo importance-sampling method of Sec. 2.2.2. Let us consider a system of N particles and write a microstate, $\bar{\Omega}$, as $\bar{\Omega} = (m_1, m_2, \ldots, m_N) = \{m_i\}_{i=1}^N$. In order to describe the most general case to be encountered in the applications, we shall to each single-particle state, m, associate an internal degeneracy, D_m. The canonical density function, Eq. (2.1.1), is then modified accordingly

$$\rho(\{m_i\}_{i=1}^N) = Z^{-1}(\prod_{j=1}^N D_{m_j})\exp[-H(\{m_i\}_{i=1}^N)/k_B T], \tag{2.2.12}$$

where Z is the generalized partition function

$$Z = \sum_{\{m\}} (\prod_{j=1}^N D_{m_j})\exp[-H(\{m_i\}_{i=1}^N)/k_B T]. \tag{2.2.13}$$

For the sake of simplicity, we choose $\bar{\bar{P}}^*$ of Eq. (2.2.9) corresponding to single-site transitions, i.e. $m_k \to m_k'$. (It is straightforward to generalize the description to account for any combination of single-site excitations.) For convenience, we introduce the internal energy and the internal entropy associated with the transition $\bar{\Omega}' = (m_1, m_2, \ldots, m_k, \ldots, m_N) \to \bar{\Omega}'' = (m_1, m_2, \ldots, m_k', \ldots, m_N)$

$$\Delta E = H(\bar{\Omega}'') - H(\bar{\Omega}') \tag{2.2.14}$$

$$\Delta s = k_B \ln(D_{m_k'}/D_{m_k}). \tag{2.2.15}$$

A possible realization of the Markov chain may then be described by the following simple algorithm:

(i)   Choose an arbitrary (e.g. random) initial configuration, $\overline{\Omega}'_1$.

(ii)   Pick a trial state, $\overline{\Omega}''_2$, according to the probability p^*_{12}.

(iii) If $\Delta E - T\Delta s \leq 0$, i.e. $D_{m'_k}/D_{m_k} \exp(-\Delta E/k_B T) \geq 1$, the trial state is nominated as the next element in the chain, $\overline{\Omega}'_2 = \overline{\Omega}''_2$.

(iv)   If $\Delta E - T\Delta s > 0$, i.e. $\lambda \equiv D_{m'_k}/D_{m_k} \exp(-\Delta E/k_B T) < 1$, a random number,[*] $\zeta \in [0, 1]$, is drawn. If $\lambda > \zeta$, the trial state is nominated as the next element in the chain, $\overline{\Omega}'_2 = \overline{\Omega}''_2$. If $\lambda \leq \zeta$, the original state is duplicated, $\overline{\Omega}'_2 = \overline{\Omega}'_1$.

(v)   A new trial state, $\overline{\Omega}''_3$, is considered — etc.

In the above algoritm, we have assumed that the thermodynamic temperature is positive. The realization is readily transcribed to apply to negative temperatures (Sec. 5.2.5, Mouritsen and Knak Jensen 1978).

If $\overline{\overline{P}}$ fulfils the ergodicity requirement, the procedure (i) - (v) will, in the limit of a large number of trial moves (*Monte Carlo steps*), lead to a distribution of states given by the canonical distribution function, Eq. (2.2.12). This limit distribution of states constitutes the equilibrium ensemble at temperature T for the model under consideration.

$\overline{\overline{P}}^*$ is in most cases chosen so as to make the sampling procedure as fast as possible. A popular choice for lattice models corresponds to a sequential or random visitation of lattice sites and to the simplest possible single-site excitation. For a model with q discrete single-site states (Ising and Potts models), this choice may be a uniformly random choice among the q possible states. For a model with a continuous degree of freedom, e.g. characterized by a unit vector, the simplest possible excitation corresponds to choosing a new uniformly random direction on the unit hypersphere. All these types of excitation may be called Glauber-like excitations since they resemble the dynamics of the kinetic Ising model (Glauber 1963). Variations of the Glauber technique could be collective excitations of particles on several sites. Use of Glauber-like excitations leads to dynamics with non-conserved order parameters. Another possible mechanism of excitation follows Kawasaki-like dynamics (Kawasaki 1972) which conserves the order parameter. A simple Kawasaki mechanism is two-site exchange of single-site properties. Obvious combinations of Glauber and Kawasaki dynamics immediately suggest themselves. With these two very simple methods of excitation, it is often possible to construct, for a given problem, a mechanism which to a very high degree mimics the real physical excitations.

---

[*] The quality of pseudo-random numbers (obtained from standard random number generators) required for good Monte Carlo work is discussed by Binder (1984).

2.2.4   General limitations of the Monte Carlo method

The realization of the Monte Carlo method described in the preceeding section allows a numerically exact calculation of Eq. (2.1.3). Of course, the exact calculation is not feasible in reality since only finite Markov chains ($M < \infty$) can be constructed. Moreover, only a finite number of particles N can be studied. The usefulness of the whole approach rests on the empirical fact that it has turned out to be possible by this Monte Carlo method to simulate fairly accurately the thermodynamic properties of macroscopic systems ($N \sim 10^{23}$) using finite systems ($N \sim 10^2 - 10^5$) and finite Markov chains ($M/N \sim 10^2 - 10^4$). The feasibility of the method has been tested on systems with non-trivial Hamiltonians which allow an exact analytical calculation of Eq. (2.1.3) (see e.g. Landau (1976a) for a Monte Carlo study of the two-dimensional Ising model in zero field).

Practical limitations therefore imply that Monte Carlo simulation of thermodynamic behavior is based on approximations involving finite values of N and M. It is important to stress, however, that these approximations are of a controlled type and that their influence on the simulation results may be determined by a systematic study using a series of values for M and N. The related convergence problems and finite-size effects will be dealt with in Secs. 2.2.7 and 2.2.8, respectively.

In order to approximate the thermodynamic limit ($N \rightarrow \infty$), one often makes the finite systems studied periodically infinite by imposing suitable boundary conditions, e.g. toroidal boundary conditions. The boundary conditions establish translational invariance and eliminate surface effects. A periodically infinite system will resemble an infinite system as long as the correlation length does not exceed the linear dimension of the system. Since the correlation length diverges at a critical point, finite-size effects become particular troublesome when critical fluctuations set in (cf. Sec. 2.2.8). The boundary conditions impose some serious restrictions on simulations of systems with ordered structures. Obviously, a finite lattice can only accomodate structures which are commensurate with the linear extension of the lattice. The disadvantage of finite-size and surface effects may be turned into an advantage when problems in surface and interface physics are at issue (see e.g. Landau 1979b). In particular, Monte Carlo simulation seems ideal to study the properties of small clusters of particles (Müller-Krumbhaar 1979).

In order to make practicable the realization of the Monte Carlo method as described in Sec. 2.2.3, it is essential that the internal energy of a given microstate is easy to evaluate. This will be the case for most classical systems with short-range interactions. It is possible, though, to do Monte Carlo calculations on quantum systems (Handscomb 1963, Suzuki 1976, Kolb 1983) and models with long-range interactions (Levesque et al. 1984). However, the complexity of such simulations has so far precluded detailed studies of phase transitions and critical phenomena.

2.2.5   Broken ergodicity

The realization of the Monte Carlo method described in Sec. 2.2.3 will provide thermodynamic equilibrium provided that the stochastic matrices $\overline{\overline{P}}$ and $\overline{\overline{P}}^{*}$ are chosen to be ergodic. Ergodicity requires that any one configuration of the Markov chain is accessible from any other configuration via a finite number of transitions

$$\forall i,j \ \exists n : \ p_{ij}^{(n)} > 0. \tag{2.2.16}$$

Equation (2.2.16) may be fulfilled by simply choosing the one-step transition probabilities, p_{ij}, of Eq. (2.2.4) such that $\forall i,j : p_{ij} > 0$. Whether this is possible or not depends on the Hamiltonian, H. If H diverges in a finite region of phase space (which will be the case of hard-sphere potentials), the Markov chain may in some cases be reducible and split up into several ergodic classes (see e.g. Wood 1968). For all applications considered in this work, the Hamiltonians are finite bounded functions and such complications will not arise.

Nevertheless, effectively broken ergodicity will occur and it has to in systems undergoing phase transitions! In fact, broken ergodicity may signal a phase transition. To understand why this is the case, it is useful to recall that any simulation as well as any experiment is associated with a time scale which is determined by the duration of the observation, τ_{obs}. If relaxation processes exist on time scales $\tau_r \gg \tau_{obs}$, the experiment will not describe the true canonical equilibrium and an effective ergodicity-breaking may be observed. To be specific, let us consider a magnetic system undergoing a phase transition to an ordered phase characterized by an order parameter Φ. In the absence of symmetry-breaking fields, $-\Phi$ would be an equally good order parameter. Furthermore, the order parameter may have several degenerate components, $\pm\Phi_1, \pm\Phi_2, \ldots$, which are connected by the operations of the symmetry group of the ordered structure. Below the phase transition, the system is equally likely to order in any one of its order parameter components. Let τ_r denote the time scale on which transitions among the components take place. In the thermodynamic limit, the canonical description leads to the ergodic phase space average

$$\forall T : \ \lim_{N\to\infty} \lim_{t\to\infty} \Phi(t, N) = 0 \tag{2.2.17}$$

which is in obvious conflict with experimental observations made on the time scale τ_{obs}. This conflict arises because the limit, Eq. (2.2.17), is not appropriate for real experiments with finite observation times $\tau_{obs} < \infty$, to which rather the limit

$$\forall T < T_c : \ \lim_{t\to\tau_{obs}\ll\tau_r} \lim_{N\to\infty} \Phi(t, N) = \pm\Phi(T) \tag{2.2.18}$$

applies. This in turn is caused by the fact that the free energy barriers between the states characterized by different order parameter components diverge at least as fast as $N^{(d-1)/d}$ where d is the spatial dimension (Palmer 1982). The relaxation time is then $\tau_r \sim \exp(aN^{(d-1)/d})$. Thus, the ergodicity is broken and the canonical description fails to describe the observation even in the thermodynamic limit.

In a computer experiment, the situation is even more complicated. Not only is the observation of the system confined to finite times, but the system itself is of a finite size, $N < \infty$. Thus, a very intricate interplay between ergodic and nonergodic behavior is expected to occur depending on the values of τ_{obs} and N chosen for a particular system. From a theoretical point of view, this is a very unusual and fortunate situation since it allows a determination of the various components of the order parameter and thus permits a characterization of the nature of the ergodicity-breaking (symmetry-breaking in this case). In fact, computer simulation may be the only way to accomplish this. This is a remarkable advantage over theoretical calculations which, in order to calculate the order parameter, require a modification of the canonical description so as to avoid the ergodic assumption of the equal *a priori* probability of all states. This can only be done by assuming the nature of the symmetry-breaking and build it into the formalism, e.g. by introducing an ordering field which is made to vanish after the thermodynamic limit is taken. Alternatively, a restricted-trace formalism may be used (Palmer 1982). In mean-field and Landau theories, a symmetry-breaking order parameter is introduced from the very beginning.

It is useful to introduce the notion of components of phase space which correspond to the components of the order parameter (Palmer 1982). Each of these components remains ergodic on a time scale τ_{obs}. Thus, during the time of observation, the system is confined to one component of phase space. By choosing appropriate values of τ_{obs} and N, one may by computer simulation not only study the effectiveness of this confinement but also observe the transitions between the various components. For a fixed temperature below the phase transition temperature, the system, which in a computer simulation is initiated in one of its components ($\overline{\Omega}_1$ of the Markov chain in Sec. 2.2.3), will be confined to that component for $t \ll \tau_r$. Decrease of N will enhance the cumulative probability of transition to another ordered component within the observation time τ_{obs}. For a fixed system size, N, this cumulative probability will increase as the temperature approaches the transition temperature. At the transition $\tau_r \rightarrow 0$, and problems may arise in interpreting the simulation data.

In computer simulations of systems with broken ergodicity, the thermal equilibrium averages are determined as averages over a single component of phase space. It can be shown that these averages are equivalent to the averages which can be derived from the complete canonical distribution including a symmetry-breaking field which is put to zero after the thermodynamic limit is taken (Palmer 1982). This equivalence extends to quantities such as $< H >$ and Φ, but not to the corresponding fluctuation quantities, C_h and χ_T, which differ by an amount of the order of N^{-1} in the restricted and unrestricted ensembles. The difference is due to intercomponent fluctuations. Rather than estimating this difference, computer simulations are usually set up, by appropriate choices of τ_{obs} and N, to make the difference negligible.

For the systems we are going to consider in Chaps. 2 - 5, the order parameter almost invariably has at least the trivial degeneracy, $\pm\Phi$. Account of possible transitions between various components of the phase space is therefore usually taken by calculating

$$\Phi \equiv < |M| >, \qquad (2.2.19)$$

where M is the order parameter operator. If the system during the simulation only spends a negligible time in passing the barriers between the different components, Eq. (2.2.19) will constitute a good approximation to the equilibrium value of the order parameter. However, the closer the system is to T_c, the more time will be spent in excursions between the components, and the more will Eq. (2.2.19) be an underestimate of the actual long-range order. To help determine Φ in those cases, the distribution functions described in the following section will be useful.

2.2.6 Distribution functions

When there is a strong interaction between the various components of the order parameter in a phase with partially broken ergodicity, it becomes difficult to estimate the order parameter in a computer simulation. This will always be the case close to continuous transitions and near first-order phase transitions influenced by fluctuations. In those cases, the order parameter distribution function, $P(\Phi)$, is helpful. Furthermore, $P(\Phi)$ is indispensible when the order parameter is used to determine the nature of a phase transition (cf. Sec. 2.2.9).

$P(\Phi)\,d\Phi$ is defined as the probability that the order parameter will take on a value in the range $[\Phi, \Phi + d\Phi]$. To be specific, let us consider a system with a one-component order parameter which is subject to the trivial degeneration, $\pm\Phi$ (The extension to systems with more than one order parameter component is straightforward.). A schematic drawing of $P(\Phi)$ is shown in Fig. 2.2.1 for a series of different temperatures relative to a phase transition temperature, T_c. $P(\Phi)$ in this figure is the complete distribution function for a finite system and

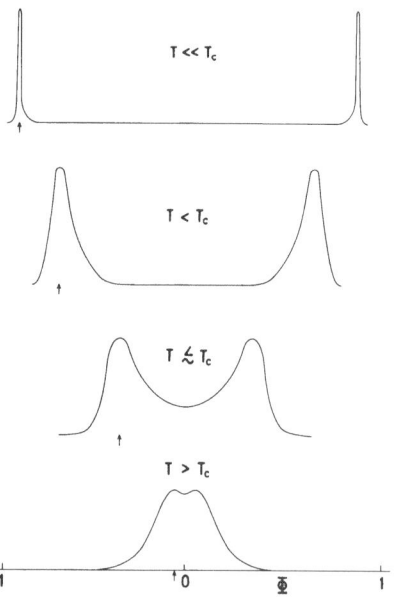

Fig. 2.2.1. Schematic drawing of the distribution function, $P(\Phi)$, of the order parameter for a finite system with a one-component order parameter. Results are shown for different temperatures relative to a phase transition temperature, T_c. The functions are normalized arbitrarily. The arrows indicate the approximate equilibrium values of the order parameter.

it is constructed using information about both components of phase space. $P(\Phi)$ has a double-peaked structure even above the transition where the finite size of the system induces a finite long-range order (cf. Sec. 2.2.8). The observable equilibrium properties of the system in the symmetry-broken phase are calculated using only one of the peaks. Due to interaction between the two components, the peaks are not perfect gaussians but have a tail extending towards $\Phi = 0$. It is this tail which, as $T \rightarrow T_c$, makes Eq. (2.2.19) less and less accurate in estimating the equilibrium order parameter. A much more accurate estimate is obtained from the position of the peaks in $P(\Phi)$.

Similarly, it is found that the distribution function for the internal energy, $P(E)$, also deviates from a gaussian shape and that the equilibrium internal energy may be estimated from the position of the maximum of the distribution. Obviously, distribution functions may be useful for other quantities, too.

An extremely interesting application of distribution functions for *local* order parameters for finite blocks of spins in hypercubic Ising models has been reported by Binder (1981a). These distribution functions constitute the basis for a Monte Carlo coarse-graining and renormalization group calculation which leads to very accurate values of critical exponents. In particular, the joint distribution function of nearest-neighbor block order parameters has proved useful in the first step towards determining the phenomenological Ginzburg-Landau Hamiltonian for specific microscopic models (Kaski et al. 1983a).

2.2.7 Coarse-graining techniques and criteria of convergence

In this section, we discuss the determination of thermodynamic equilibrium averages from Monte Carlo simulations assuming that the system is effectively confined to a single component of phase space. If this assumption fails, a combination of the techniques described in this section and those described in Sec. 2.2.6 has to be applied.

In principle, the limit distribution of the Markov chain of microconfigurations generated by the Monte Carlo method is independent of the initial configuration. In practice, only finite Markov chains can be obtained and correlations in the chain will to some extent reflect the choice of initial state. The question then arises as to how large M needs to be in order to provide a reliable approximation to the equilibrium ensemble. A useful procedure by which this question may be answered proceeds via coarse-grained averages. Coarse-grained averages are averages over subsets of the Markov chain, $\{\overline{\Omega}'_1, \overline{\Omega}'_2, \ldots, \overline{\Omega}'_M\}$, cf. Sec. 2.2.2. Each subset consists of ΔM configurations. The nth coarse-grained average of a quantity f is defined as

$$<f>_n = \Delta M^{-1} \sum_{j=(n-1)\Delta M+1}^{n\Delta M} f(\overline{\Omega}'_j). \qquad (2.2.20)$$

The following function is then studied

$$F_m = (m+1)^{-1} \sum_{j=0}^{m} <f>_{M/\Delta M - j} \qquad (2.2.21)$$

for various values $m < M/\Delta M$. If F_m attains a reasonably stable value over an extended number of coarse-grained averages, we consider this value to be an estimate of the equilibrium average, $<f>$. The reverse summation in Eq. (2.2.21) enables correlations in the Markov chain to be revealed. In particular, the partial sums facilitate a determination of which of the initial coarse-grained averages should be discarded. The minimum value of M required to determine $<f>$ with a given accuracy depends on how far the temperature is from the transition temperature as well as on the property f itself. The most demanding properties are fluctuation quantities. Close to a critical point, the relaxation towards equilibrium is extremely slow (»critical slowing down«) and at T_c the relaxation time diverges (see e.g. Binder and Kalos 1979). Thus, special care should be taken to avoid trapping in metastable states.

From the experience with computer simulations on a great variety of different systems undergoing phase transitions, it emerges that the most reliable determination of convergence properties is based on fairly subjective criteria. Use of these criteria involve »eye balling« of large amounts of »raw« data for coarse-grained averages of many different properties. It also involves the study of the evolution of the Markov chain using many different initial configurations as well as different random number sequences.

In the choice of the stochastic matrix, $\overline{\overline{P}}^*$, of Eq. (2.2.9), there is plenty of room for an »experimental« optimization of the convergence rates. There is no *a priori* choice of $\overline{\overline{P}}^*$ which is known to optimize the convergence. The choosing of $\overline{\overline{P}}^*$ is made solely on the basis of physical intuition and by trial and error.

2.2.8 Finite-size effects

The most serious drawback of a computer simulation approach to the study of phase transitions is that one must deal with finite systems. No finite system with a non-singular Hamiltonian can exhibit a true phase transition. This becomes obvious when noting that the partition function in Eq. (2.1.2) cannot develop a singularity when the integral of the finite bounded Boltzmann function is performed over a finite phase space. Nevertheless, finite systems have reminiscences of phase transitions, and systematic studies of these pseudo-transitions as functions of system size may reveal information about the phase transition in the infinite system.

A finite system gives an accurate description of the infinite system as long as the correlation length, ξ, does not exceed the linear extension, L, of the system. When $\xi \gtrsim L$ (e.g. near a critical point), the properties of the finite system will reflect the nature of the boundary conditions. For systems with periodic boundary conditions, the fluctuations will be »over-correlated« and the various properties will be »rounded«. As an example, the long-range order will persist above the phase transition and singularities in the specific heat and the ordering susceptibility will be rounded and shifted in temperature.

Ferdinand and Fisher (1969; see also Fisher 1971) have developed a finite-size scaling theory for critical phenomena[*] which is extremely useful to guide the extrapolation of Monte Carlo finite-system properties to the thermodynamic limit. According to this theory, the free energy of a finite system is given by the homogeneous function

$$F(N, T) = L^{-(2-\alpha)/\nu} \bar{F}(tL^{1/\nu}), \qquad (2.2.22)$$

where α and ν are the critical exponents pertaining to the specific heat and to the correlation length, respectively. $\bar{F}$ is a scaling function involving the scaled variable $tL^{1/\nu}$ only, and $t = (T - T_c)/T_c$ with $T_c = T_c(L = \infty)$.

Fisher (1971) has suggested the position of the maximum of the specific heat (or alternatively the ordering susceptibility) as an appropriate definition of the »transition temperature« of the finite system, $T_c(L)$. According to the finite-size scaling theory, the shift in critical temperature then scales as

$$\delta T_c = T_c(L = \infty) - T_c(L) \sim L^{-1/\nu}. \qquad (2.2.23)$$

From the free energy in Eq. (2.2.22), the scaling properties of other thermodynamic functions may be derived, e.g.

$$\Phi = L^{-\beta/\nu} \bar{M}(tL^{1/\nu}). \qquad (2.2.24)$$

In the limit, $t \to 0$ and $L \to \infty$, Eq. (2.2.24) has to reduce to the infinite-system singular behavior, Eq. (2.1.9), and therefore, in this limit, the order parameter scaling function is given by

$$\bar{M}(tL^{1/\nu}) \sim (-tL^{1/\nu})^\beta. \qquad (2.2.25)$$

The various scaling functions and amplitudes are non-universal properties which depend on the details of the system under consideration. In particular, these non-universal properties are functions of the boundary conditions chosen.

The validity of the finite-size scaling theory has been demonstrated through extensive Monte Carlo simulations on two- and three-dimensional Ising models with periodic boundary conditions as well as free surfaces (Landau 1976a,b). It appears from these calculations that systems with $L \gtrsim 10$ are well inside the asymptotic region described by Eq. (2.2.22). Therefore, corrections-to-finite-size scaling, e.g.

$$\delta T_c \sim L^{-1/\nu}(1 + aL^{-\Delta} + \dots), \qquad (2.2.26)$$

where Δ is the correction-to-scaling exponent ($\Delta \sim \frac{1}{2}$) (cf. Eq. (2.1.13)), need not be invoked. This is a very important result because it anticipates that Monte Carlo simulations of critical phenomena are feasible (cf. Sec. 3.1.4) using system sizes and statistics compatible with current computer capacities.

---

[*] The finite-size scaling theory for first-order phase transitions (Fisher and Berker 1982) has so far not been used to analyze Monte Carlo simulation data.

2.2.9   Determining the nature of a phase transition

Elucidating the nature of phase transitions by computer simulation techniques is hampered by finite-size effects. A phase transition is infinitely sharp only in an infinite system. The characteristic discontinuities and singularities accompanying phase transitions will appear as rounded and smeared in the finite systems employed in computer simulations. In real systems, phase transitions will also appear smeared to a degree which depends on the size of the system and which especially depends on the concentration of imperfections and impurities.

In principle, it is impossible, by any laboratory experiment or computer simulation, to prove that a phase transition is continuous. It can always be postulated that possible first-order discontinuities are below the resolution of the experiment. Similarly, it may be argued that experimentally observed metastabilities do not signal first-order phase transitions but are non-equilibrium effects associated with continuous transitions. The best one can do is to analyze as many properties as possible in the neighborhood of the phase transition and compare them with phase transitions of a well-established nature. Thus, laboratory and computer experiments share a deficiency in their inability to determine unambiguously the nature of phase transitions.

The very first question to be answered in a computer simulation study of the cooperative behavior of an interacting many-body system is related to the existence of a stable ordered phase at finite temperatures. Theoretically, this is known to be a very difficult problem (see e.g. Griffiths 1972). In computer simulation studies, this probem is approached by a finite-size analysis of the possible long-range order parameters. The possible types of ordering in the finite system are suggested by the simulation itself. If the order parameter for finite temperatures approaches a non-zero value as $L \rightarrow \infty$, it is concluded that a phase with long-range order remains stable at finite temperatures and that therefore a finite-temperature phase transition must exist (cf. Sec. 5.3.4). *

Having established the existence and structure of the ordered phase, we may now turn to the question of the nature of the phase transition. The transition is triggered when a thermodynamic parameter, e.g. the temperature, is varied. For the sake of simplicity, we shall restrict ourselves to situations with a single phase transition. Examples of systems with multiple phase transitions are discussed in Secs. 5.2.6 and 5.2.8. As described in Sec. 2.2.2, we shall distinguish between first-order and continuous phase transitions only.

A first-order phase transition is characterized by a discontinuity in the order parameter. The specific heat has a δ-function singularity superimposed on its discontinuity at T_c. The energy content of the δ-function represents the enthalpy of the transition. An important indication of a first-order phase transition is the presence of metastable states in the transition region. In this region, the free energy has two minima and the system may become trapped in the upper metastable

---

* Evidence for the absence of conventional long-range order in the square antiferromagnetic Potts model and the two-dimensional classical Heisenberg model has recently been delivered by Monte Carlo renormalization group transformations on pair correlation functions (Shenker and Tobochnik 1980, Jayaprakash and Tobochnik 1982).

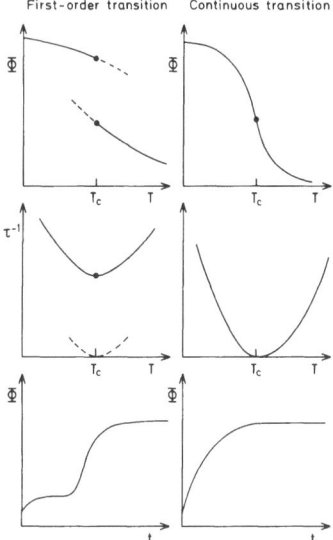

First-order transition   Continuous transition

Fig. 2.2.2. Variation of order parameter, $\Phi(T)$, and inverse relaxation time, $\tau^{-1}(T)$, with temperature, and variation of order parameter with Markov time, t. Results are shown for first-order and continuous transitions in the case of a finite system. T_c is the transition temperature. The dashed lines signify the region of metastable states associated with the first-order transition. The inverse relaxation time has two branches for the first-order transition, an upper one corresponding to the stable states and a lower one corresponding to the metastable states. [Adapted from Landau and Binder 1978].

one for a finite time $t < \tau(T)$, where $\tau(T)$ is the relaxation time of the metastable state. In Fig. 2.2.2 is shown schematically the variation with temperature of the order parameter, $\Phi(T)$, and the inverse relaxation time in the region around a first-order transition. The metastable branches of $\Phi(T)$ are indicated. The end points of these branches are pseudospinodal points at which the metastable system becomes thermodynamically unstable. The inverse relaxation time, accordingly, has two branches. The relaxation out of the metastable states becomes infinitely slow as the equilibrium transition point is approached from either side. When the system is perturbed in a metastable state, e.g. by an appropriate change in temperature, it may exhibit a two-step relaxation behavior which is an exceptional feature of systems undergoing first-order transitions: Firstly, the system relaxes into a new metastable state characterized by the new temperature and then eventually it relaxes into the true equilibrium state. The important point to stress is that the second step of the relaxation may not set in for a long time and that therefore in some cases the metastable state may mistakenly be confused with the equilibrium state. The macroscopic consequence of trapping in metastable states is the observability of hysteresis, i.e. the behavior of the system in the transition region depends on its thermal history. There are a few important exceptions to make from the above description. Firstly, some systems may have several local minima of the free energy in the transition region and may therefore give rise to a more complicated pattern of metastable states. This may in some cases (cf.

e.g. Fig. 5.1.6) lead to a cascading relaxation behavior associated with a whole staircase of steps. Secondly, the presence of hysteresis may not necessarily signal a first-order transition. If the system under consideration has more order parameter components than physical dimensions, an extremely slow domain-growth kinetics may result when the system is taken below the transition point (cf. Sec. 5.4). In that case, the annealing of domains characterized by different order parameters will be the rate-determining process and in many cases it is impossible within a reasonable observation time to bring the system into a uniformly ordered phase. A global hysteresis will then result (cf. Figs. 5.3.9 and 5.3.10) irrespective of the specific nature of the phase transition.

At a continuous transition, the order parameter vanishes in a continuous manner, cf. Eq. (2.1.9), and the fluctuation quantities may diverge, cf. Eqs. (2.1.10) and (2.1.11). The free energy only has a single minimum and no metastabilities are expected. However, the relaxation to equilibrium becomes slower as T_c is approached and the relaxation time diverges at the critical point (»critical slowing down«). Thus, very close to T_c and for sufficiently short observation times, a system with a continuous transition may behave as being effectively in a metastable state. The temperature variation of $\Phi(T)$ and $\tau^{-1}(T)$ is drawn schematically in Fig. 2.2.2. Note that this figure applies for a finite system which supports long-range ordering even at and above T_c. Figure 2.2.2 also shows the time evolution of the order parameter. Close to T_c, a plateau does not develop until a very long time has elapsed.

In computer simulations on finite systems, first-order phase transitions will appear as partly smeared: the discontinuities are reduced and the δ-function of the specific heat is broadened. For first-order transitions associated with strong fluctuations, these observations are the usual consequences of finite-size rounding (Sec. 2.2.8). However, there is always a contribution to the smearing from the effective averaging of information from metastable and stable states. Since a finite system can give rise to only a finite free energy barrier between the two minima of the free energy function, there is a finite probability for crossing the barrier within the observation time. This crossing is not only allowed from the local (metastable) minimum to the global (stable) minimum, but a finite system may perform a shifting between the two. If the barrier is very low (which it will be close to T_c), this will lead to a complete smearing and it may be difficult to resolve the two states. In that case, the first-order transition effectively appears as a continuous transition strongly dominated by fluctuations. However, by increasing the lattice size, the two minima may be resolved and the complete free energy surface may be accurately probed by the simulation. This is a remarkable advantage of computer simulation over conventional theoretical calculations which are usually only able to probe the equilibrium properties. Detection of the metastable states is facilitated by studying the time-evolution of coarse-grained averages and the structure of distribution functions of the internal energy and the order parameter. The distribution functions are particularly useful if the system has several order parameter components. In that case, the shifting between ordered and disordered states near T_c is coupled with a shifting between the various components of the order parameter. If the shifting between the stable and metastable states is sufficiently rapid to sample

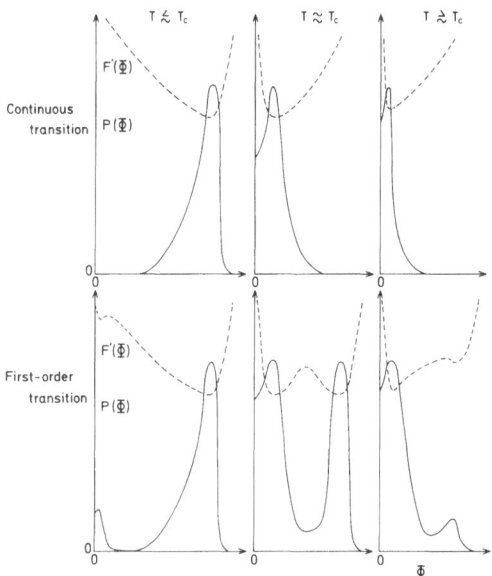

Fig. 2.2.3. Distribution functions, $P(\Phi)$ (solid lines), and the corresponding free energy functional, $F'(\Phi)$ Eq. (2.2.27) (dashed lines), for a system with a one-component order parameter at temperatures immediately below, at, and immediately above the equilibrium phase transition temperature, T_c. The upper panel corresponds to a continuous transition and the lower panel to a first-order transition. The drawing is purely schematic. All curves are displayed on arbitrary ordinate scales and the same common arbitrary absissa scale.

accurately the complete distribution function for, say, the order parameter, $P(\Phi)$, a unique way is offered for determining the equilibrium first-order phase transition temperature. This is illustrated in Fig. 2.2.3 which shows $P(\Phi)$ in the transition region of a finite system with a single order parameter component. (For systems with the trivial $\pm\Phi$ degeneracy, the complete $P(\Phi)$ is obtained from Fig. 2.2.3 by a reflection in the $\Phi = 0$ line. The extension to systems with more order parameter components is straightforward.) $P(\Phi)$ is a double-peaked function throughout the transition region where metastable states persist. The figure illustrates that, at $T \simeq T_c$, the two peaks have the same intensity. This indicates that both phases are equally likely and that we are therefore at the only point where the two phases can coexist, i.e. at the equilibrium transition point. To be consistent with a first order transition, it is an important requirement that the two peaks in Fig. 2.2.3 move apart as the system size is increased.[*] Glosli and Plischke (1983) have recently pointed out that $P(\Phi)$ is related to a free energy functional

$$F'(\Phi) = -k_B T \ln P(\Phi),\qquad(2.2.27)$$

which is part of the total free energy. The equivalent use of $F'(\Phi)$ to analyze the nature of the phase transition is perhaps somewhat more heuristic. $F'(\Phi)$ is also shown in Fig. 2.2.3.

---

[*] See Landau and Swendsen (1981) for a situation where the opposite size dependence of the splitting in $P(\Phi)$ indicates a continuous transition.

If the complete distribution function, $P(\Phi)$, cannot be obtained for systems with first-order transitions, a few other methods may be called upon to determine T_c. The first one uses the classic Maxwell equal-area rule. This is not a very accurate method since it presupposes, incorrectly, that the relaxation time is symmetric about T_c. A second method, which has found wide use among high-energy physicists to locate phase transitions in lattice gauge theories (see e.g. Creutz et al. 1979), is a mixed-phase calculation. The idea is to initiate the simulation by a configuration which is a one-to-one mixture of the two phases. The equilibrium phase is then determined as the phase of the mixture which grows as the ensemble is built up. This method requires less computer time but is not as accurate as the method which uses the complete $P(\Phi)$ function. The third method requires an evaluation of the free energy function itself. Since $F(T)$ is not a thermal average, its evaluation presupposes knowledge of the partition function, Z. However, Z is not available from a Monte Carlo calculation which is built on Eq. (2.2.9). Therefore, $F(T)$ has to be determined indirectly,[*] e.g. by a numerical integration of the internal energy

$$F(T) = \frac{T}{T_i} F(T_i) + T \int_{T_i^{-1}}^{T^{-1}} E(x)\, dx. \tag{2.2.28}$$

Such a procedure requires detailed information on the internal energy over a wide range of temperatures around the transition as well as precise knowledge of appropriate high- and low-temperature boundaries, $F(T_i)$, where the integration is started. (Applications of this procedure will be described in Secs. 3.2.3 and 5.1.8.) The free energy is then calculated for each phase separately. If the internal energy is known along the metastable branches of the hysteresis loop, the point at which the two resulting free energy branches intersect may be determined fairly accurately. This point is the equilibrium transition point. From the difference in slope of the two $F(T)$ branches in this point, the enthalpy and entropy of transition may be derived

$$\Delta S/T_c = \Delta E = \Delta(\partial F(T)/\partial T)_{T_c}. \tag{2.2.29}$$

The distribution function, $P(\Phi)$, for a finite system with one order parameter component undergoing a continuous transition is also shown in Fig. 2.2.3. $P(\Phi)$ has a single peak for all temperatures and the position of the peak moves continuously as the temperature is varied through the phase transition. The corresponding free energy functional, $F'(\Phi)$ Eq. (2.2.27), has a single minimum. These characteristics are distinctly different from those found for first-order transitions. The finite-size behavior of the various thermodynamic functions are conveniently analyzed in terms of the finite-size scaling theory presented in Sec. 2.2.8 to yield critical temperature and exponents.

---

[*] See Binder (1979) for a list of references to other procedures of evaluating $F(T)$.

2.2.10 Computational details

Good Monte Carlo work requires large amounts of computer time, often of the order of 10 - 100 cpu hours on standard general-purpose computers. Thus, highly optimized and careful programming is required for the central loops in the simulation programs. The requirement of a fast code is to some extent in conflict with the need to deal with a large number of particles. The characteristics of the individual particles as well as the local topology of the system preferably have to be stored in the central memory since this information is accessed randomly. Bit-manipulation and the use of logic operations may often reduce requirements made to memory and time, especially for systems with discrete degrees of freedom. Methods also exist by which several sites can be activated simultaneously (Jacobs and Rebbi 1981).

The astoundingly rapid development of computer technology makes it redundant to describe in detail how Monte Carlo techniques are implemented on current computers. We shall rather outline some significant trends in the application of modern computers to statistical mechanics. For the average Monte Carlo computer experimentalist, the general-purpose mainframe computers and the proliferating minicomputers will probably continue to be the most popular forms of hardware. However, since most problems in statistical mechanics require little data handling and much data processing, microcomputers may soon gain ground, especially for problems which can be formulated in terms of discrete variables (see e.g. Bak (1983) for a microcomputer study of the critical behavior of the three-dimensional Ising model).

Vector and array processors are bound to gain an enormous impact on the field of simulation since they allow parallel processing and simultaneous Monte Carlo activation on several sites. This will particularly be true of special general-purpose array processors, such as the Edinburgh Distributed Array Processor which is constructed from 2^{12} interlinked microprocessors (Hirsch and Scalapino 1983, Pawley et al. 1983). The most recent development is the invention of special-purpose computers which are »hard wired« to deal with a specific statistical mechanical problem. Examples are the Delft (Hooghland et al. 1983) and Santa Barbara Processors (Pearson et al. 1983). Special-purpose computers possess the advantages of being very cheap and, once being constructed, they may be devoted full-time to a given problem.

To illustrate the potentials of these new developments on the hard-ware side of computational physics, we shall quote typical performances of the various computers in terms of the cpu time required to make a successful Monte Carlo spin excitation in the three-dimensional nearest-neighbor Ising model: CYBER 173 and VAX 11/780: $\gtrsim 10^{-4}$sec; CDC 7600: $\gtrsim 10^{-6}$sec; Delft Processor: $\sim 7 \times 10^{-7}$sec; CRAY 1: $\sim 5 \times 10^{-7}$sec; Edinburgh Distributed Array Processor: $\sim 2 \times 10^{-7}$sec; and the Santa Barbara Processor: $\sim 4 \times 10^{-8}$sec. The system sizes which can currently be treated in the special-purpose computers are for example $\leq 128^3$ spins in the Santa Barbara Processor and $\leq 2^{22}$ spins in the Delft Processor (Hirsch and Scalapino 1983). This should be compared with the sizes of the largest systems which may realistically be studied on general-purpose computers. These sizes are

180^3 for studies of critical phenomena and up to 800^3 for studies of percolation (Stauffer 1982). However, in contrast to the special-purpose computers, the largest systems studied by general-purpose computers partly have to be stored in auxiliary memory devices.

For the scientist with easy access to a general-purpose computer or, perhaps, a personal microcomputer, Monte Carlo simulation of cooperatively behaving many-body systems will fairly soon become a standard technique. As it appears from the description of the conventional Monte Carlo method in the present chapter, the method and its realization are very simple and flexible and the scientist need not be a computer specialist to carry out useful simulation studies on her favorite model of nature.

2.2.11 General advantages of the Monte Carlo method: Applications

The drawbacks of the Monte Carlo method as described in Sec. 2.2.4 are amply outbalanced by its general advantages, such as simplicity, universal applicability, and its susceptibility to systematic improvement.

There is a remarkable duality associated with the use of numerical computer simulation methods when studying many-body systems. On the one hand, a substantial amount of data processing is involved which might be thought to obscure the physical meaning of the results. On the other hand, the computer is not a »black box« as is most complicated experimental equipment. Rather, every step of the simulation is transparent. The availability of the microscopic states allows a close-up picture to be formed of the static as well as the dynamic behavior of the system. The ability to »watch« a system on the molecular level greatly enhances the »feeling« for the important physics.

Basically any thermodynamic quantity may be determined, ranging from bulk thermodynamic properties and response functions to microscopic correlation functions. Furthermore, insight into time-dependent processes may be gained via the dynamical interpretation of the Monte Carlo method.

By numerical simulation, it is often possible to overcome some of the mathematical difficulties associated with analytical theoretical calculations in statistical mechanics. From a theoretical point of view, there are two general advantages of major importance within computer simulation. First of all, it is a non-perturbative approach and secondly, computer simulation is an unbiased approach which only presupposes the fundamental physical laws of nature. Therefore, computer simulation facilitates a determination of the relevant variables of a problem and may eventually lead to new and unexpected physical discoveries. Compared to conventional theoretical calculations, computer simulations have the following special advantages: they operate by controlled approximations only, they can easily treat complicated Hamiltonians, and they require no additional approximations when new features are added to the Hamiltonian.

It is important to emphasize that Monte Carlo calculations may be considered numerical »experiments« (computer experiments) and they are therefore subject to many of the same advantages and drawbacks as are laboratory experiments. Still, compared to laboratory experiments, numerical experiments have a number of assets: they are applied to clean and well-defined systems, they are performed under well-controlled conditions, they may be carried out under extreme physical conditions, and they provide detailed information on the molecular level.

The various applications of computer simulation in statistical mechanics may conveniently be divided into the three main classes given in the Introduction. The potentials of computer simulation mirrored in these different kinds of application put the field of computer simulation in a remarkable position relative to theory and experiment. It plays the important mediating role in the dialectic interplay between testing and proposing models of nature and thus it touches the foundations of scientific reasoning.

2.3 Exact Configurational Counting and Series Expansions

2.3.1 A general approach

For more than two decades, analysis of series expansions has provided some of the most significant and reliable information on the critical properties of non-trivial lattice models (for a review, see Domb and Green 1974). This information has been of seminal importance in proposing, testing, and corroborating the basic ideas of the scaling and universality hypotheses. Results from series analysis are often considered to be the ultimate judge in assessing predictions of approximate theories for critical phenomena, e.g. the renormalization group theory.

Series expansion techniques are usually characterized by their specificity to the model and lattice under consideration. Highly specialized techniques are required to obtain the large number of expansion coefficients which is necessary to determine the finer details of critical behavior. Therefore, only fairly simple models with a few model parameters have been subjected to series analysis.

Recently, a general scheme has been proposed to calculate high-temperature series expansions for general spin Hamiltonians containing a large number of model parameters (Mouritsen 1979, Mouritsen et al. 1980).* The scheme yields the series coefficients as exact functions of the model parameters, e.g. spatial dimensionality, coupling distribution in coordinate and spin space, site-dependent field distributions, and spin quantum number.

Here, the principles of this general approach will be described. We shall pay particular attention to the philosophy behind the method because it represents an interesting and unconventional computer study of phase transitions and critical phenomena. Contrary to the *numerical* computer simulation procedures described in Sec. 2.2, the present computer approach is purely *analytical*. *Throughout*, the calculation of the series coefficients via this new scheme is carried out utilizing a computer. Indeed, without the aid of a fast computer this general approach would

---

* Other rather general schemes include the linked cluster expansion (Wortis 1974) and the direct graphical expansion for Ising systems (Oitmaa 1981).

not be feasible. When deriving the series expansions, the computer essentially acts as a »bookkeeper«: Topological objects, or *graphs*, are generated and compared, combinatorial factors associated with the graphs are determined, and the number of possible embeddings of the various graphs in a physical object, the lattice, is calculated. All operations done by the computer are of a purely symbolic character. The final results are *exact* analytical expressions in the chosen variables which have an integer representation in the computer.

The use of a computer to solve combinatorial problems provides many advantages of which the most important — apart from a high speed of calculation — is the possibility to construct a reliable machine program which can be applied with only minor modifications to different Hamiltonians. However, utilizing a computer also creates problems (Martin 1974) of which the most serious in the present implementation is the use of labeled graphs corresponding to a specific bit representation in the computer. Working sequentially on labeled graphs excludes extensive use of graph theoretical symmetry arguments in calculating the weights derived from the topology of the graphs. This implies that every graph contributing to a specific coefficient must be generated and compared piece by piece with the already existing graphs so as to determine all distinct graphs and the associated weights or multiplicities. To illustrate this point, we may mention that a typical sixth-order term for the cases we are considering contains several millions of graphs of which only a few hundred are found to be different. For comparison, the number of distinct interaction graphs for the isotropic Heisenberg model is of the order of ten (Rushbrooke et al. 1974). It should be noted that the general approach, as well as all other series expansion techniques applied to many-body problems, is characterized by the unpleasant feature that the work necessary to generate the individual coefficients of the series is tremendous and very complicated. Furthermore, this work increases progressively with the order making it a highly specialized field. Finally, the calculations of the series coefficients in itself provides almost no physical insight into the underlying problem! This is in marked contrast to computer simulations which can provide extremely useful information while the statistics for the equilibrium ensemble is built up.

The major advantage of the general approach is that the series expansions for many different models can be derived from the same general expression. The drawback is that fewer series coefficients can be obtained than if more specific methods were applied. This implies that information on the critical properties extracted from the series is less accurate. Our general series extend to *sixth order* in the inverse temperature. The strength of the general approach lies in the calculation of phase diagrams of complex models for which series analysis has not previously been practicable.

In Secs. 2.3.2 – 2.3.5, an account of the general principles of the technique will be given. The description serves to reveal the capabilities of the approach as well as its limitations.

2.3.2   The moment method

The *moment method* constitutes a convenient basis for a computer-based general approach to calculate high-temperature series expansions. The moment method,[*] or the Kramers-Opechowski method, was originally introduced by Opechowski (1937) to derive high-temperature series expansions for the spin$-\frac{1}{2}$ isotropic Heisenberg model.

We consider a system of N equivalent spins with spin quantum number S and governed by the Hamiltonian H. In thermal equilibrium at a temperature T, the system is described by the density operator, Eq. (2.1.1),

$$\rho(T) = e^{-H/k_BT}/Tr(e^{-H/k_BT}). \qquad (2.3.1)$$

The statistical mean value of a quantum mechanical observable represented by the spin operator F is expanded in a Taylor series around the non-interacting limit $T = \infty$

$$<F> \equiv Tr[\rho(T)F]$$

$$= \sum_{n=0}^{\infty} \frac{1}{n!}(\partial^n <F>/\partial K^n)_{K=0} K^n$$

$$= \sum_{n=0}^{\infty} \frac{(-1)^n}{n!} c_n K^n. \qquad (2.3.2)$$

In Eq. (2.3.2), we have used the high-temperature expansion variable $K \equiv J/k_BT$, where J is an energy parameter. In the case $F = H$, the coefficients c_n are related to the averages $\{<H^m>\}_{m=1}^n$ as are the cumulants to moments in the theory of statistics, c_n being the semiinvariant average with H as the relevant statistical variable (Kirkwood 1938, Horwitz and Callen 1961). The generalization to $F \neq H$ is straightforward leading to

$$c_0 = \ll F \gg$$
$$c_1 = \ll FH \gg - \ll F \gg \ll H \gg$$
$$c_2 = \ll FH^2 \gg - 2 \ll FH \gg \ll H \gg - \ll F \gg \ll H^2 \gg + 2 \ll F \gg \ll H \gg^2$$
$$\cdots \qquad (2.3.3)$$

with the following definition of the reduced trace

$$\ll \cdot \gg \equiv Tr(\cdot)/(2S + 1)^N. \qquad (2.3.4)$$

Eq. (2.3.2) should properly be termed a »low-K-« rather than a »high-T« expansion as it applies also to the cases of negative temperatures (Sec. 5.2).

---

[*] We refer to the articles in the book by Domb and Green (1974) for a general reference as to how the moment method relates to other current methods, e.g. the cumulant method, the finite cluster method, and the linked cluster expansion.

We shall here restrict ourselves to systems with short-range forces. If we ignore boundary effects and consider the N-spin system as being mechanically homogeneous, we can argue on physical grounds that each cumulant in Eq. (2.3.3) is of the same order in N as is the quantity $< F >$. For an extensive thermodynamic quantity proportional to N, the only part of c_n which contains terms of order N is $<< FH^n >>$. The remaining terms of $<< FH^n >>$, which are of higher orders in N, are bound to cancel the remaining part of the cumulant. Thus, the essential problem in calculating high-temperature series coefficients via the moment method is reduced to calculating quantities like $Tr(FH^n)$. In order to compute the nth order coefficient, we have to determine all connected graphs corresponding to $Tr(FH^n)$ and to evaluate the traces associated with each graph.

2.3.3   Principles of the calculation [*]

The general approach is capable of treating Hamiltonians of the form

$$H = \sum_{j=1}^{N} \sum_{i=1}^{M} \sum_{u=1}^{L(i)} \prod_{k=1}^{l_i} O_{s_k}^{(q)} \prod_{m,n \in \{s_1, s_2, \ldots, s_{l_i}\}} B_{mn}^{(t)}, \tag{2.3.5}$$

where $O_{s_k}^{(q)}$ is a spin operator of type $q = q(k, i)$ belonging to the s_kth spin. $O_{s_k}^{(q)}$ may in the general case be a product of powers of the components of the spin operator $\bar{S}_{s_k}$. The first sum in Eq. (2.3.5) runs over all N spins of the system and for each spin, j, the second sum runs over M interaction topologies specifying the types of interaction in which the jth spin participates. The ith topology is characterized by l_i index names, $\{s_1, s_2, \ldots, s_{l_i}\}$, denoting the spins with operators $O_{s_k}^{(q)}$ involved in the interaction corresponding to the topology. The set $\{s_1, s_2, \ldots, s_{l_i}\}$ includes j itself. Finally, the ith topology is associated with a coupling distribution $B_{mn}^{(t)}$ of type t. The third sum in Eq. (2.3.5) runs over $L(i)$ realizations of the ith topology. A realization gives the actual values of the index names $\{s_1, s_2, \ldots, s_{l_i}\}$.

A class of Hamiltonians is defined as the subset of H in Eq. (2.3.5) resulting from a specific choice of the interaction topologies. It is obvious that Eq. (2.3.5) includes as special cases most Hamiltonians normally used for lattice spin models, e.g. pair (exchange, dipolar, Dzyaloshinsky-Moriya) and multi-spin interactions with various symmetry-breaking homogeneous or inhomogeneous external fields (or crystal fields) included. For each class, the lattice structure (including spatial dimensionality), the distribution of coupling and field parameters in coordinate and spin space (including range of interaction), and the spin quantum number appear as variables.

We now turn to the calculation of $Tr(FH^n)$, where F is a spin operator of type $O^{(q)}$ which may contain fixed indices (e.g. in the case of two-point correlation functions). A general term of the trace may be written as

---

[*] The presentation follows very closely that of Mouritsen (1979).

$$Tr(\sum_{k_1,k_2,\ldots,k_n} B_{i_1j_1}^{(1)} B_{i_2j_2}^{(2)} \ldots B_{i_lj_l}^{(l)} O_{k_1}^{(1)} O_{k_2}^{(2)} \ldots O_{k_m}^{(m)})$$

$$= \sum_{k_1,\ldots,k_n} B_{i_1j_1}^{(1)} B_{i_2j_2}^{(2)} \ldots B_{i_lj_l}^{(l)} Tr(O_{k_1}^{(1)} O_{k_2}^{(2)} \ldots O_{k_m}^{(m)}) \tag{2.3.6}$$

with the restriction

$$\{i_1,j_1, i_2,j_2, \ldots, i_l,j_l\} \subseteq \{k_1, k_2, \ldots, k_m\}. \tag{2.3.7}$$

The set $\{k_{n+1}, k_{n+2}, \ldots, k_m\}$ constitutes the fixed indices. The inclusion sign in Eq. (2.3.7) indicates that each index *name* must appear at a spin operator. Specifying the class of Hamiltonians under consideration is equivalent to giving the actual correspondence between the two sets of index names in Eq. (2.3.7).

Each term in Eq. (2.3.6) may be mapped on a linear graph (Domb 1974), where the indices (or the traces of spin operators $O_j^{(p)}$) correspond to the vertices and the coupling parameters, $B_{ij}^{(t)}$, to the lines (bonds). The calculation procedure of Eq. (2.3.6) may conveniently be divided into two basic steps:

Step 1. Determination of all distinct connected graphs and their multiplicities.

Step 2. Embedding of connected graphs into a lattice.

Step 1 is fairly general and may be adapted to various homogeneous many-body systems (solid, liquid, gas) with different degrees of freedom. The series coefficients resulting from step 1 contain a significant amount of configurational information of a general character, and it is important to realize that up to this point the result only depends on the interaction topologies chosen. In step 2, we introduce the spatial configuration of the system (i.e. lattice structure and dimension). This step may also be termed calculation of »lattice constants« (Domb 1974). In the approach of Domb and collaborators, both steps are performed simultaneously in order to reduce the number of contributing graphs by taking into account the specific nature of the Hamiltonian and the lattice under consideration. For the present purpose, we would gain no advantage by incorporating the embedding procedure in step 1 because practically all distinct connected graphs resulting from step 1 can be embedded into the lattice due to the large number of interacting neighbors in the Hamiltonian. Nevertheless, we shall impose some restrictions on step 1, of which the symmetry of the Hamiltonian and the properties of the spin angular momentum operators make the most significant reduction in the number of contributing graphs.

2.3.4 Step 1. Determination of all distinct graphs and their multiplicities

The starting point is Eqs. (2.3.6) and (2.3.7) which implies an N^n-fold summation. Even for a simple interaction topology and to a low order this is a tremendous and tedious task. Fortunately, only a limited number of terms are non-zero and only a small number of distinct graphs contributes. In the case of a simple Hamiltonian (e.g. the spin $-\frac{1}{2}$ Ising model in zero field), the few distinct graphs and their multiplicities may be determined by hand to a high order, but for the more complex cases which we want to consider this is not manageable beyond second or third order. Therefore, the computer is »taught« to take care of all the necessary »bookkeeping« involved in the summation. This type of analytical computer calculation is a generalization of the procedure introduced in the derivation of the higher moments of the magnetic resonance lines of a dipolar coupled rigid lattice (Knak Jensen and Kjaersgaard Hansen 1973).

The sum in Eq. (2.3.6) is an unrestricted sum, i.e. all summation indices take on all possible values independently. In order to calculate the traces, we transform this unrestricted sum into a sum of restricted sums $\sum'$:

$$\underset{k_1, k_2, \ldots, k_n}{\sum{}'} \equiv \sum_{k_1} \; \underset{k_2 (\neq k_1)}{\sum} \; \cdots \; \underset{k_n (\neq k_1, k_2, \ldots, k_{n-1})}{\sum} . \qquad (2.3.8)$$

The sums $\sum'$ are restricted in the sense that no pair of summation indices take on the same value simultaneously. Rewriting the right hand side of Eq. (2.3.6) as

$$\underset{k_1, k_2, \ldots, k_n}{\sum} P(k_1, k_2, \ldots, k_m) = \underset{k_1, k_2, \ldots, k_n}{\sum} E(i_1, j_1, \ldots, i_l, j_l) Tr(\prod_{i=1}^{m} O_{k_i}^{(i)}) \qquad (2.3.9)$$

with the restriction Eq. (2.3.7), the transformation is given by

$$\underset{k_1, k_2, \ldots, k_n}{\sum} P(k_1, k_2, \ldots, k_m) = \sum_G \; \underset{i_1, i_2, \ldots, i_{u_G}}{\sum{}'} P^G(i_1, i_2, \ldots, i_{u_G}). \qquad (2.3.10)$$

In the Gth restricted sum in Eq. (2.3.10), the indices $k_1, k_2, \ldots, k_m$ are divided into u_G groups and all indices in a group are substituted by a single new index. $P^G(i_1, i_2, \ldots, i_{u_G})$ is obtained by making this substitution in $P(k_1, k_2, \ldots, k_m)$. In other words, $\sum_G$ is a sum of all groupings of summation indices, each grouping being a disjunctive coverage of the set $\{k_1, k_2, \ldots, k_m\}$. If fixed indices are contained in the set, as in the case of F being a two-point correlation function, their names are kept fixed in order to identify these at the end of the calculation. Using the transformation, Eq. (2.3.10), in Eq. (2.3.9), we obtain

$$\underset{k_1, k_2, \ldots, k_n}{\sum} E(i_1, j_1, \ldots, i_l, j_l) Tr(\prod_{i=1}^{m} O_{k_i}^{(i)})$$

$$= \sum_G \; \underset{i_1, i_2, \ldots, i_{u_G}}{\sum{}'} E^G(i_1, i_2, \ldots, i_{u_G}) Tr(\prod_{p=1}^{u_G} O_{i_p}^{(G,p)})$$

$$= \sum_G (2S+1)^{N-u_G} (\prod_{p=1}^{u_G} tr(O^{(G,p)}))(\underset{i_1, i_2, \ldots, i_{u_G}}{\sum{}'} E^G(i_1, i_2, \ldots, i_{u_G})).$$

$$(2.3.11)$$

All spin operators with index i_p are collected in the operator $O^{(G,p)}$. The trace $tr(O^{(G,p)})$ is defined on the space of a single spin variable. Equation (2.3.11) is the basic equation of step 1. In evaluating Eq. (2.3.11), we use the property of the spin angular momentum operators that $Tr[(S_x)^{n_x}(S_y)^{n_y}(S_z)^{n_z}]$ is zero unless n_x, n_y, and n_z are all even or all odd integers (Ambler et al. 1962).

An alternative formulation of the calculation involved in step 1 may be given in the context of graph theory. It is very instructive to make a graph representation of the arithmetic process involved in Eqs. (2.3.6) through (2.3.11), but we emphasize that our approach is purely algebraic from a conceptual point of view and makes no use of graph theoretical theorems or simplifications. This is strongly coupled with the basic distinction between labeled an unlabeled graphs (Domb 1974). In using a computer, we necessarily have to work with labeled graphs. The procedure in going from Eq. (2.3.6) to Eq. (2.3.11) corresponds in graph theoretical language to the determination of all distinct graphs (and their multiplicities) which can be constructed from the m vertices, $\{k_1, k_2, \ldots, k_m\}$, and the l lines $\{B_{i_1 j_1}^{(1)}, B_{i_2 j_2}^{(2)}, \ldots, B_{i_l j_l}^{(l)}\}$. Both the vertices and the lines may overlap in the resulting graphs which can be either connected or disconnected (Domb 1974). Each vertex corresponds to a trace of spin operators. In the construction of the different traces, the non-commutativity of the spin operators must be taken into account. The computer program determines *all* graphs and sums up the number of identical graphs in order to determine their multiplicities. This implies that the calculation of the multiplicities by the present method consumes the major part of the computation time in contrast to other current methods (Domb 1974, Rushbrooke et al. 1974) which often determine these numbers by a simple combinatorial formula.

The graphs resulting from Eq. (2.3.11) are either connected or disconnected. Each graph, which corresponds to a restricted sum, is termed a restricted graph. Due to the linked-cluster property of the cumulants in Eq. (2.3.3), we have to consider only connected graphs. If the quantity under consideration is of order N^p, a disconnected restricted graph contains terms of the order N^p and $N^{p+1}, N^{p+2}, \ldots, N^{p+n}$, where $n + 1$ is the number of disconnected constituents (subgraphs) of the graph. This entails that the disconnected graphs - in their capacity of being restricted - contain a connected part which must be isolated. This final procedure in step 1 is also programmed for a computer and the output contains a sum of terms each characterized by a connected graph and a linear combination of single-spin traces. The values of the necessary traces are given by Ambler et al. (1962) or may easily be evaluated algebraically on a computer (Knak Jensen and Kjaersgaard Hansen 1973). All quantum mechanical traces are polynominals in $S(S + 1)$.

2.3.5   Step 2. Embedding of connected graphs into a lattice

This step assumes a specific underlying lattice into which the graphs are to be embedded or, alternatively, the lattice constants are to be evaluated. The distribution of coupling parameters, B, now has to be specified including the number of interacting neighbors and the range of interaction. In the case of a large number of interacting neighbors with different strength of interaction, the embedding problem is almost as time-consuming as the determination of the multiplicities of the graphs in step 1. The calculation may in some cases be substantially reduced by taking into account the spatial symmetry of the distribution B. In the case of calculating properties with site-dependent weight factors (e.g. the staggering index for the ordering susceptibility of antiferromagnets), only the common symmetry of the distribution B and the weight factors can be exploited.

The embedding procedure is also programmed for a computer. We refer to Martin (1974) for a detailed description of computer techniques for evaluating lattice constants. In the present implementation, the calculation is straightforward: Fix one vertex at some site of the lattice and vary the rest of the vertices on the lattice sites with the restrictions imposed through B. Because we deal exclusively with restricted sums (or graphs), the enumeration of the embeddings is closely related to the excluded volume problem.

In closing the formal description of the general approach to a computer-based derivation of high-temperature series expansions, we want to stress the extreme importance attached to a thorough checking of the computer code. Obviously, the codes for carrying through steps 1 and 2 are significantly more complicated than the most sophisticated Monte Carlo simulation program. Slight errors may build up and influence the final series coefficients. Even small numerical errors in the higher-order coefficients may fatally distort the asymptotic analysis of the series. The computer programs for the general approach have been checked in various limiting cases by comparison with series for well-known models reported in the literature (Mouritsen 1979). Once a reliable computer program is devised, we possess the superior advantage of having at our disposal a machinery which can readily be applied to new problems without further checking procedures being required.

2.3.6   General correlation function series

A basic quantity of a spin system is the spin-spin correlation function, $\Gamma_{\alpha\beta}(\bar{r}_{ij}, T)$, which is a measure of the degree of correlation between spin fluctuations at lattice sites i and j. For $T > T_c$ and in zero external field, $\Gamma_{\alpha\beta}$ takes the following form for systems with no off-diagonal interactions

$$\Gamma_{\alpha\beta}(\bar{r}_{ij}, T) \equiv <(S_{i\alpha} - <S_{i\alpha}>)(S_{j\beta} - <S_{j\beta}>)> / Tr(S_\alpha^2)$$
$$= \delta_{\alpha\beta} <S_{i\alpha}S_{j\alpha}> / Tr(S_\alpha^2). \qquad (2.3.12)$$

$S_{i\alpha}$ is one of the components of the spin vector $\bar{S}_i = (S_{ix}, S_{iy}, S_{iz})$.

From the correlation function, a variety of physical quantities can be constructed, e.g.: (i) For systems with pair interactions, the internal energy is simply given as a linear combination of pair correlations within the interaction sphere. From the internal energy, the specific heat can be derived. (ii) According to the fluctuation theorem, the wave vector-dependent susceptibility tensor $\chi_\alpha(\bar{q}, T)$ is the Fourier transform of $\Gamma_{\alpha\beta}$. (iii) From the wave vector-dependent spherical moments

$$\mu_{\alpha,n}(\bar{q}, T) \equiv \sum_{j(\neq i)} (|\bar{r}_{ij}|/r_0)^n \, e^{i\bar{q}\cdot\bar{r}_{ij}} \Gamma_{\alpha\alpha}(\bar{r}_{ij}, T), \qquad (2.3.13)$$

the correlation length may be determined as

$$\xi_\alpha(\bar{q}, T) = r_0[\mu_{\alpha,2}(\bar{q}, T)/2d\mu_{\alpha,0}(\bar{q}, T)]^{\frac{1}{2}}, \qquad (2.3.14)$$

where r_0 is the lattice constant and d is the spatial dimension. For $T \to T_c$, the correlation length defined from the second spherical moment is expected to be proportional to the true correlation length (Fisher and Burford 1967)

$$\xi_\alpha(\bar{q}, T) \equiv \lim_{|\bar{r}_{ij}| \to \infty} r_0|\bar{r}_{ij}|^{-1} \ln|e^{i\bar{q}\cdot\bar{r}_{ij}} \Gamma_{\alpha\alpha}(\bar{r}_{ij}, T)|. \qquad (2.3.15)$$

The general approach constitutes a convenient method for calculating correlation function series. The availability of auto and pair correlation function series as functions of $\bar{r}_{ij}$ allows studies of several thermodynamic quantities which are expected to display critical fluctuations, e.g.

$$\mu_{\alpha,n}(\bar{q} = \bar{q}_0, T) \sim (T - T_c)^{-\gamma - n\nu}, \quad T \to T_c \qquad (2.3.16)$$

$$\mu_{\alpha,0}(\bar{q} = \bar{q}_0, T) = k_B T N^{-1} \chi_\alpha(\bar{q}_0, T) - \Gamma_{\alpha\alpha}(\bar{0}, T) \qquad (2.3.17)$$

$$\simeq k_B T N^{-1} \chi_\alpha(\bar{q}_0, T), \quad T \to T_c$$

$$\sim (T - T_c)^{-\gamma}, \quad T \to T_c \qquad (2.3.18)$$

$$\xi_\alpha(\bar{q} = \bar{q}_0, T) \sim (T - T_c)^{-\nu}, \quad T \to T_c, \qquad (2.3.19)$$

where $\bar{q}_0$ is the wave vector characterizing the ordered phase.

There exists a number of techniques to analyze finite series for critical behavior. These techniques have been reviewed extensively by Hunter and Baker (1973, 1979, Baker and Hunter 1973), Gaunt and Guttmann (1974), and by Pearce (1978). The series obtained from the general approach are rather short and they allow only analysis in terms of the leading singularities, e.g. Eqs. (2.3.16) – (2.3.19). For that purpose only rather simple methods of asymptotic analysis are required, such as ratio, Neville, and Padé analyses as well as critical point renormalization. To isolate the physical singularities of interest and to improve the convergence, various conformal mappings are usually also applied to the series.

2.3.7 Capabilities and limitations of a general approach

The general approach described above is capable of treating rather general spin Hamiltonians. The practical limitations of the approach are set exclusively by the capacity of the computer utilized. For most series expansion techniques, there exists a rather sharp upper limit as to the number of series coefficient which may in practice be determined. For the present approach, this limit corresponds to the sixth order in the inverse temperature. The associated computer times are of the order of several days. Presumably, the fastest current computers may add an extra term to the series in roughly the same computer time. The general nature of the approach makes it insuperably difficult to reach beyond this. We suggest that further progress may be obtained by »teaching« the computer how to recognize general symmetries of the interaction graphs occuring in step 1 by parallel processing of indices.

So far, the general approach has mainly been used to derive phase diagrams of classical three-dimensional antiferromagnetic spin models characterized by anisotropic distributions of couplings in spin and coordinate space and subject to symmetry-breaking fields. Among these are the truncated secular dipolar model of nuclear spin ordering (Sec. 5.2), the antiferromagnetic anisotropic Heisenberg model in a field (Mouritsen et al. 1980), and certain antiferromagnetic models with many-component order parameters (Sec. 4.1). Because the series extend only to sixth order, it is impossible to determine critical parameters with the high accuracy that characterizes series analyses based on model-specific expansion techniques. However, the large number of model parameters which can be included in the series coefficients makes it possible to determine phase diagrams and critical behavior for a large class of models for which series analysis has not previously been practicable.

In Secs. 4.1 and 5.2 will be given the results of high-temperature series analysis of a variety of phase diagrams for antiferromagnetic models. Several of these models have also been investigated by numerical computer simulation techniques. It will be demonstrated that the composite results of series analyses and computer simulations give an exhaustive description of complex antiferromagnetic phase diagrams.

3. Monte Carlo Pure-model Calculations

Within statistical mechanics it is considered a very worthwhile occupation to study simplified and pure microscopic models of interacting many-body systems. These models are often constructed with the sole purpose of determining the minimum set of model features which are necessary to produce a given physical phenomenon, e.g. a phase transition. The simplicity of the models serves to make transparent the precise relationship between the mathematical structure of the model and its thermodynamic properties. In many cases, the original construction of the models has been governed by mathematical tractability rather than physical realism.

The motivations for the theorist's preoccupation with pure and simple models are several. Firstly, pure models have a considerable interest in their own right. Secondly, as far as critical phenomena are concerned, the concept of universality provides a justification for a study of simple and pure models as being representatives of the various universality classes. Finally, a large number of models first studied for mathematical convenience have later — often to the originators' surprise — turned out to be excellent models for specific physical materials and phenomena. Indeed, a great deal of truth is buried in the theorist's popular saying that no matter how peculiar a model she studies, sooner or later the experimentalist will discover a real physical system or phenomenon which the model may describe.

The focus in this chapter will be on the two most celebrated pure models studied in statistical mechanics, the *Ising model* and the *Heisenberg model*. Both of these models were originally suggested as simplified lattice models of ferromagnetism. In its lattice gas representation, the Ising model also found an early application as a model of alloys. Introduced by Lenz in 1920 (Lenz 1920), solved in one dimension by Ising in 1925 (Ising 1925) and in two dimensions by Onsager in 1944 (Onsager 1944), the »simple Ising model still thrives« (Fisher 1981) in the 1980s. In the present context, we shall speak of Ising models as a class of statistical mechanical lattice models defined in terms of finitely discrete commuting local scalar variables. The Hamiltonian of the models contains simple products of these variables. An immense literature is available on Ising models, and over the years these models have been proposed to describe a diversity of systems ranging from magnetic materials, alloys, and polymers to biological membranes and the human brain.

The Ising model with short-range pair interactions does not give rise to a finite-temperature phase transition in less than two spatial dimensions. In two dimensions, the exact Onsager-solution predicts critical behavior. Despite its simplicity, the three-dimensional Ising model has so far escaped an exact solution. Thus, our knowledge about the phase behavior of this model is rooted in approximate analytical calculations, series expansions, real-space renormalization group calculations, and computer simulations. In Sec. 3.1, it will be demonstrated how Monte Carlo simulations can be used to study the phase transition in the spin $-\frac{1}{2}$ Ising model with nearest-neighbor pair interactions on a three-dimensional lattice. That section serves as the »textbook« example of the present survey and illustrates the main potentials and limitations of conventional Monte Carlo work in critical phenomena. The presentation is aimed at the non-specialist and the more experienced reader is suggested to proceed to Sec. 3.2. In that section we discuss the effects on the phase behavior of Ising models due to the presence of multi-spin interactions.

The second pure model studied by computer methods in this chapter, the Heisenberg model, is characterized by three-component spin vectors coupled by a pair interaction Hamiltonian which contains scalar products of these vectors. The Heisenberg model supports conventional long-range order only when the spatial dimension is larger than two. In Sec. 3.3, we shall study classical (i.e. non-quantum) *one-dimensional* Heisenberg chains with anisotropic pair interactions subject to external magnetic fields. Although such models do not have finite-temperature phase transitions, they merit study because their thermodynamics is influenced by collective non-linear excitations (solitons).

3.1 Critical Behavior of the Three-dimensional Ising Model

3.1.1   The Ising model and its order parameter

The spin $-\frac{1}{2}$ Ising model with isotropic nearest-neighbor ferromagnetic pair interactions is defined by the Hamiltonian

$$H = -J \sum_{<i,j>} \sigma_i \sigma_j, \quad J > 0. \tag{3.1.1}$$

The model is arrayed on a lattice with N sites and each site is assigned a spin $-\frac{1}{2}$ variable, $\sigma_i = \pm 1$. With this definition, the model has a simple $n = 1$-component order parameter, the spontaneous ferromagnetic magnetization,

$$m(T) = N^{-1} < | \sum_{i=1}^{N} \sigma_i | > . \tag{3.1.2}$$

For a sufficiently large lattice and not to close to the transition temperature T_c, the ergodicity of the model is effectively broken in the ordered phase (cf. Sec. 2.2.5) and transitions are very unlikely between the two states globally magnetized »up« and »down«. The low-temperature thermodynamics is then derived within a subspace of the total phase space as described in Sec. 2.2.7.

An extensive Monte Carlo study of the two-dimensional quadratic Ising model has been reported by Landau (1976a). The results of this study as well as of previous Monte Carlo studies (see. e.g. Yang 1963, Fosdick 1963, Friedberg and Cameron 1970, Stoll et al. 1973) have, when compared with the exact solution, provided a strong confidence in a computer-based approach to critical phenomena in Ising models. Landau (1976b) also studied the simple cubic Ising lattice by Monte Carlo techniques (for earlier studies, see e.g. Fosdick 1963, Binder 1972, Binder and Hohenberg 1974; see also Sec. 4.2.2).[*] Here, we shall describe the results of a Monte Carlo study of the ferromagnetic Ising model on a non-Bravais lattice structure, the diamond lattice.

3.1.2 Numerical evidence of a phase transition in the Ising model on a diamond lattice

The diamond lattice is represented by four equivalent interpenetrating sublattices. Each sublattice is a tetragonally distorted simple cubic lattice of cubical form with L^3 lattice points subject to toroidal periodic boundary conditions. The total number of lattice points is then $N = 4L^3$. The Hamiltonian is that given in Eq. (3.1.1) and the range of interaction is restricted to the four nearest neighbors of the diamond lattice.

A Monte Carlo study of this model has been reported by Mouritsen (1980) who employed the computational principles described in Sec. 2.2. The calculations are performed on lattices with linear dimensions $L = 4, 6, 9, 12$, and 16 corresponding to spin systems with N ranging from 256 to 16384. The results are based on statistics involving 2000 – 5000 Monte Carlo steps per site, depending on the temperature.

Figure 3.1.1 shows the temperature dependence of the nearest-neighbor and the next-nearest-neighbor correlation functions, $< \sigma_0 \sigma_1 >$ and $< \sigma_1 \sigma_2 >$. The former is proportional to the internal energy per spin, $E(T) = < H > /N = -2J < \sigma_0 \sigma_1 >$. Pronounced finite-size effects are seen in the correlation functions. Both functions have an inflection point close to $k_B T/J \simeq 2.7$. The slope of the tangent in the inflection point increases with increasing lattice size. Within statistical error, the data for the two larger lattices coalesce indicating that the thermodynamic limit has been reached effectively. In going from $< \sigma_0 \sigma_1 >$ to $< \sigma_1 \sigma_2 >$, a stronger temperature dependence is observed for the pair correlation function. In the limit of large separations $r \to \infty$, $< \sigma\sigma(r) >$ tends towards $m^2(T)$. The variation with temperature of the order parameter, $m(T)$ Eq. (3.1.1), is shown in Fig. 3.1.2. A strong dependence on the lattice size is observed above $k_B T/J \simeq 2.65$ whereas below this limit the order parameter is fairly insensitive to the lattice size. Only in the low-temperature region, has the thermodynamic limit been reached as far as $m(T)$ is concerned. The evidence presented in Figs. 3.1.1 and 3.1.2 suggests consistently that the model undergoes a phase transition in the neighborhood of $k_B T/J \simeq 2.7$. The continuous variation of the correlation functions as well as of the order parameter indicates that the transition is continuous, in accordance with expectations. The large high-temperature tails of $m(T)$ are thus caused by the longe-range correlations which persist in a finite system even above the transition temperature.

---

[*] For a more complete list of references to Monte Carlo calculations on three-dimensional Ising models on a variety of lattices, including further neighbor interactions and magnetic fields, see e.g. Landau 1979a.

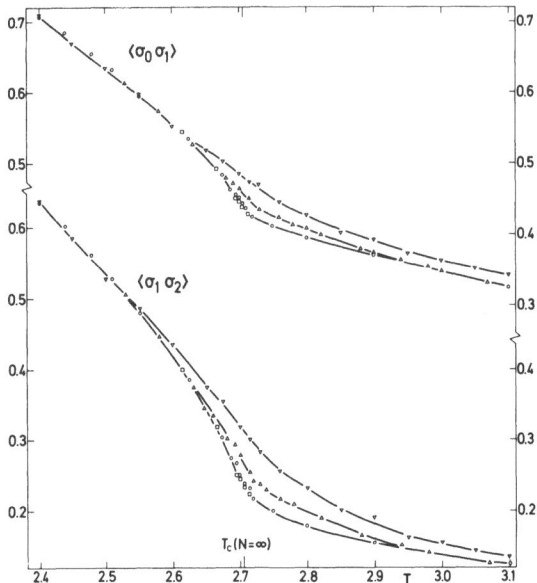

Fig. 3.1.1. Temperature dependence of the nearest-neighbor correlation function, $< \sigma_0 \sigma_1 >$, and the next-nearest-neighbor correlation function, $< \sigma_1 \sigma_2 >$. Monte Carlo data are given for four different lattice sizes: $N = 4L^3$, $L = 4$ (inverse triangles), $L = 6$ (triangles), $L = 12$ (circles), and $L = 16$ (squares). The temperature is in units of J/k_B. $T_c(N = \infty)$ denotes the critical temperature of the infinite lattice.

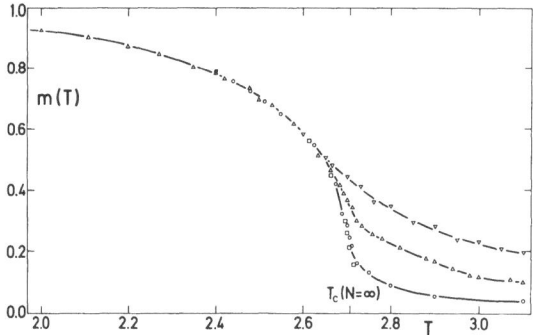

Fig. 3.1.2. Temperature dependence of the order parameter, $m(T)$ Eq. (3.1.2). The symbols are explained in Fig. 3.1.1.

Further evidence of a critical phenomenon in the model is given in Figs. 3.1.3 and 3.1.4 which show the temperature dependence of the specific heat, $C(T)$, and the ordering susceptibility $\chi(T)$. Both of these response functions are derived from the fluctuation theorem (cf. Eq. (2.1.7) for $C(T)$) which in the case of $\chi(T)$ reads

$$\chi(T) = \frac{N}{k_B T}[N^{-2} < (\sum_{i=1}^{N} \sigma_i)^2 > - m^2(T)]. \qquad (3.1.3)$$

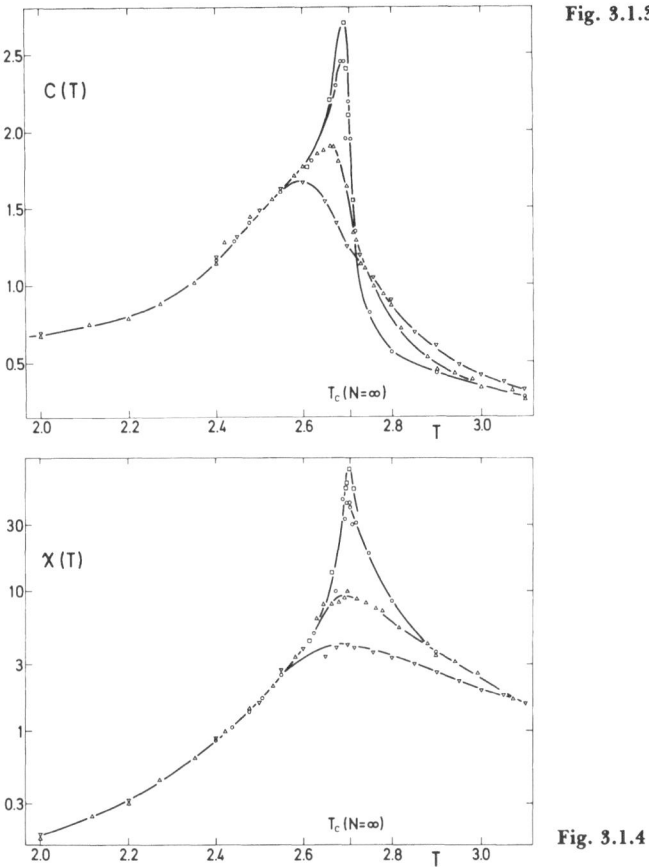

Fig. 3.1.3.

Fig. 3.1.4

Fig. 3.1.3. Temperature dependence of the specific heat, $C(T)$, in units of k_B. The symbols are explained in Fig. 3.1.1.

Fig. 3.1.4 Temperature dependence of the ordering susceptibility, $\chi(T)$, in Eq. (3.1.3). The symbols are explained in Fig. 3.1.1.

$C(T)$ as well as $\chi(T)$ exhibits a peak anomaly in the transition region. The intensity of the peaks increases with lattice size suggesting a divergence of the response functions at the transition temperature. Again, this is an indication of a critical phenomenon of the type expected for the three-dimensional Ising model.

3.1.3   Finite-size scaling analysis and critical behavior

We now proceed with a detailed quantitative analysis of the Monte Carlo data for the Ising model on the diamond lattice. The analysis aims at extracting critical parameters which can be compared with values obtained by alternative methods of calculation and which furthermore allow a discussion in the context of universality.

The critical temperature in the thermodynamic limit may be estimated using finite-size scaling theory (Fisher 1971, cf. Sec. 2.2.8). If we define the »critical temperature« of the finite system, $T_c(L)$, as the temperature where either $C(T)$ or $\chi(T)$ attains its maximum, then the shift in the transition temperature of the finite lattice relative to the infinite lattice is given by Eq. (2.2.23)

$$\delta T_c(L) = [T_c - T_c(L)]/T_c \approx aL^{-1/\nu}, \quad L \gg 1 \tag{3.1.4}$$

with $T_c \equiv T_c(L = \infty)$. a is a constant which depends on details of the model and of the nature of the boundary conditions. In Fig. 3.1.5 is presented a plot of $T_c(L)$ vs $L^{-1/\nu}$ assuming the Ising value of the correlation length exponent, $\nu = 0.630$ (Le Guillou and Zinn-Justin 1980). Considering the uncertainties involved in determining the peak positions in Figs. 3.1.3 and 3.1.4, it makes little difference whether the finite-size scaling analysis is carried out in terms of $L^{-1/\nu}$ or simply L^{-1}. First of all, it is noticed that $T_c(L)$ determined from $C(T)$ has a stronger size dependence than $T_c(L)$ determined from $\chi(T)$. This result is likely to depend on the nature of the boundary conditions (see e.g. Landau 1976a). Secondly, Fig. 3.1.5 demonstrates that the Monte Carlo data for $L \gtrsim 6$ are within the asymptotic region described by Eq. (3.1.4) with both definitions of $T_c(L)$. The smaller lattice with $L = 4$ and $T_c(L = 4) = (2.60 \pm 0.03)J/k_B$ (using the position of the $C(T)$ peak) is outside the region of validity of Eq. (3.1.4) where corrections-to-scaling are important. By a simple extrapolation to $L^{-1/\nu} = 0$ in Fig. 3.1.5, both sets of $T_c(L)$ data lead to the same estimate of the infinite-lattice critical temperature, $T_c(L = \infty) = (2.705 \pm 0.005)J/k_B$. The critical temperature for the diamond lattice has also been estimated from high-temperature series analysis of the suscep-

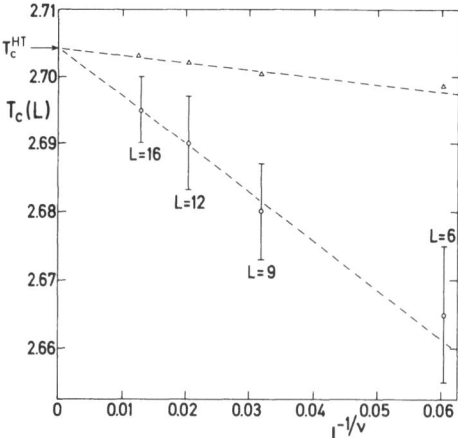

Fig. 3.1.5. Finite-size scaling plot showing the variation of the finite-lattice critical temperature, $T_c(L)$, with the linear dimension of the lattice. $\nu = 0.630$ is the critical exponent pertaining to the correlation length. Circles and triangles indicate $T_c(L)$ as estimated from the peak position in $C(T)$ and $\chi(T)$, respectively. T_c^{HT} denotes the critical temperature estimated from series analysis. The temperature is in units of J/k_B. For the sake of clarity, only error bars on the $C(T)$ data are included.

tibility (Gaunt and Sykes 1973). The result is $T_c^{HT} = (2.7043 \pm 0.0002) J/k_B$ in excellent agreement with the Monte Carlo value. The amplitude a in the scaling relation, Eq. (3.1.4), is found from Fig. 3.1.5 to be $a(diamond) = 0.27 \pm 0.05$ (using the $C(T)$ data). Landau (1976b) estimated the corresponding coefficient for the simple cubic lattice to be $a(sc) = 0.98 \pm 0.04$ with L being the linear dimension of the simple cubic lattice. If we rescale the length L pertinent to our case with $4^{\frac{1}{3}}$ in order to compare $a(diamond)$ with $a(sc)$ on the basis of the same linear lattice point density, we observe that the shift in the critical temperature due to finite-size effects is around 50% less for the diamond lattice than for the simple cubic lattice.

Other non-universal critical parameters which may be estimated from the Monte Carlo data by appropriate finite-size scaling analyses are e.g. critical pair and multi-spin correlation functions (Mouritsen 1980) which are useful in examining the validity of certain basic assumptions in approximate analytical theories for the ferromagnetic Ising transition (cf. Sec. 4.3). Here, we shall briefly mention the results of a finite-size scaling analysis of the nearest-neighbor pair correlation function, $< \sigma_0 \sigma_1 >$, for the diamond lattice to yield the critical energy, $E_c = E(T = T_c)$. E_c is of seminal importance for allowing an investigation of the confluent singularities and correction-to-scaling of the specific heat in the neighborhood of the critical point (cf. Sec. 2.1.2). Using Monte Carlo data obtained at $T = T_c^{HT}$ for a series of diamond lattices with L ranging up to 30, Knak Jensen and Mouritsen (1982) obtained an estimate of E_c using the scaling relation (Ferdinand and Fisher 1969)

$$E_c(L) - E_c(L = \infty) \sim L^{-(1-\alpha)/\nu}, \quad L \gg 1 \tag{3.1.5}$$

valid *at* the critical point. From a simple extrapolation procedure assuming the renormalization group values of the critical exponents, $E_c = E_c(L = \infty)$ was found to be $E_c = -(0.865 \pm 0.002) J$ which differs significantly from the series estimate, $E_c^{HT} = -(0.874 \pm 0.005) J$ (Hunter 1967). Conversely, for the Ising model on the simple cubic lattice, an excellent agreement was observed between the Monte Carlo and series estimates of E_c (Knak Jensen and Mouritsen 1982). Presumably, the series for the low-coordinated diamond lattice are too short to allow a realiable estimate of the critical energy.

With the critical temperature being available, we are now in a position to extract universal critical parameters, such as exponents, from the Monte Carlo data (for a Monte Carlo study of universal critical amplitude ratioes, see e.g. Knak Jensen and Mouritsen 1982 and Sec. 5.2.7). To be specific, we shall determine the order parameter exponent β from the Monte Carlo data in Fig. 3.1.2. Derivation of other exponents, such as the susceptibility exponents γ and γ' and the correlation length exponents ν and ν', from Monte Carlo data is described in Sec. 5.2.7. Finite-size scaling theory (Sec. 2.2.8) leads to the following relation for the order parameter $m(T)$

$$m(T) = L^{-\beta/\nu} \tilde{M}(x) \tag{3.1.6}$$

with the scaling variable $x = tL^{1/\nu}$. $t = [T - T_c(L = \infty)]/T_c(L = \infty)$ is the

reduced temperature, $\tilde{M}$ is a scaling function, and β is the critical exponent pertaining to the magnetization, i.e.

$$\tilde{M}(x) \simeq B(-x)^{\beta}, \quad t \ll 1, \quad L \to \infty. \tag{3.1.7}$$

B is the critical amplitude. A finite-size scaling plot of the magnetization data in Fig. 3.1.2 is presented in Fig. 3.1.6. In the plot, we have assumed the renormalization group values of the Ising ($n = 1$) exponents: $\beta = 0.325$ and $\nu = 0.630$ (Le Guillou and Zinn-Justin 1980). It appears from this plot that all the Monte Carlo data of Fig. 3.1.2 for $T < T_c$ may be described by a common function, the scaling function Eq. (3.1.6), with *no* adjustable parameter. Thus, the model obeys static scaling. Moreover, for $x \gtrsim x_0 \approx 0.6$ the data satisfy the asymptotic relation, Eq. (3.1.7), with the critical parameters

$$\beta = 0.30 \pm 0.02, \quad B = 1.63 \pm 0.04. \tag{3.1.8}$$

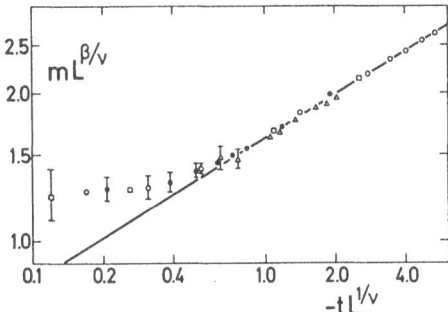

Fig. 3.1.6. Finite-size scaling plot of the order parameter, $m(T)$, according to Eq. (3.1.6). $x = tL^{1/\nu}$ is the scaling variable, L is the linear dimension of the lattice, and β and ν are the order parameter and correlation length exponents. $t = [T - T_c(L = \infty)]/T_c(L = \infty)$ is the reduced temperature. Triangles: $L = 6$, solid circles: $L = 9$, circles: $L = 12$, squares: $L = 16$. The full line represents the asymptotic behavior, Eq. (3.1.7), with the critical parameters as given in Eq. (3.1.8).

A conventional log-log plot of the $m(t)$ data leads to critical parameters consistent with Eq. (3.1.8). Furthermore, these values of the critical parameters accord with the results based on series analysis: $\beta = 0.312 \pm 0.002$ (Gaunt and Sykes 1973) and $B = 1.666 \pm 0.004$ (Domb 1974). In particular, the Monte Carlo value of β for the diamond lattice is consistent with the values of β found (by a variety of calculations, see e.g. Table 5.2.1) for the Ising model on the three cubic lattices (Domb 1974). Thus, the critical behavior of the order parameter of the Ising model on the diamond lattice is in accordance with critical behavior of the $d = 3$ $n = 1$ Ising universality class (Le Guillou and Zinn-Justin 1980).

Landau (1976b) determined the value of x_0 to be around 0.5 for the simple cubic lattice. Performing the rescaling of L described above, we obtain $x_0(diamond) \simeq \frac{1}{2}x_0(sc)$. Thus it appears that the range of the scaled variable $x = tL^{1/\nu}$, over which finite-size rounding effects of $m(t)$ are negligible, is considerably larger for the diamond lattice than for the simple cubic lattice. The smaller values of x_0

and a for the diamond lattice are considered to be due to the lower coordination number ($z = 4$) of the diamond lattice, which is likely to lead to a lower critical amplitude for the correlation length.

3.1.4 Are Monte Carlo techniques practicable in the study of critical phenomena?

After having described our »textbook example« of a conventional Monte Carlo simulation study of a simple microscopic interaction model which displays critical behavior and before we proceed to more complicated situations, it is useful to make a few reflections on the above question. We do not want to discuss the *feasibility* of the computer-based approach. This has convincingly been demonstrated by extensive calculations on models which are exactly solvable. Rather, we want to discuss the *practicability*, i.e. address questions which are related to limitations imposed by standard current computer capacities.

Two basic problems are at issue (cf. Sec. 2.2.4): (i) critical phenomena only occur in infinite systems, and (ii) the critical parameters are determined from the asymptotic behavior of functional forms only valid »close« to the critical point. The finite-size scaling assumption for continuous phase transitions (Sec. 2.2.8) combines these two problems in a single one: asymptotic critical behavior may be observed only when the scaling variable, $x = tL^{1/\nu}$, becomes »sufficiently large«, i.e. larger than some value x_o. For $x < x_o$, correction-to-scaling becomes important and rounding sets in.

The quality of the Monte Carlo data presented in Secs. 3.1.2 and 3.1.3 is representative of most conventional Monte Carlo work, i.e. the statistics ($\sim 10^3$ – 10^4 Monte Carlo steps per site) and the lattice sizes ($N \sim 10^3$ – 10^5) are as naturally dictated by most average-sized non-dedicated computers. From Fig. 3.1.6, we see that for the order parameter of the present model, »close to the critical point« means $x_o \gtrsim 0.6$. Work on other Ising models on three-dimensional lattices (Landau 1976b, 1977) indicates that x_o is of the order of unity for quantities such as order parameter and susceptibility. This is indeed a very fortunate situation because it has the important implication that, by using Monte Carlo data of average quality, the leading asymptotic form of the singular functions valid close to a critical point may be accessible over at least one decade of the scaling variable. Access to another decade in the case of the Ising model on a diamond lattice would require an increase of L by a factor of 10^ν and thus presumably involve two orders of magnitude more computer time.

The use of vector and array processors and in particular of special-purpose computers designed to study a particular Ising model, e.g. via Monte Carlo renormalization group techniques, makes practicable a much more accurate determination of critical parameters (cf. Sec. 2.2.10). As an example, the critical temperature of the simple cubic Ising model has recently been determined with a 0.002% accuracy,[*] i.e. five times more precise than even the most reliable series estimate (Adler 1983).

---

[*] Using the ICL Distributed Array Processor at Edinburgh (Pawley et al. 1983).

3.2 Phase Behavior of Ising Models with Multi-spin Interactions

3.2.1 Higher-order exchange in magnetic systems

In the theoretical description of insulating magnetic materials, it is often tacitly assumed that two-particle bilinear exchange is the only relevant interaction. Still, it has been known for a long time that in principle higher-order exchange is allowed and in many cases may play a significant role. The most important higher-order exchange terms in magnetic systems are those involving four spin operators associated with either two, three, or four particles. The sources of the fourth-order terms are several, ranging from multipole expansions of Coulomb and exchange interactions to effective spin-lattice (phonon) interactions (Aharony 1973b, 1974, Matveev and Nagaev 1972, Adler and Oitmaa 1979). In some cases, the fourth-order exchange is just as important as the bilinear exchange. Third-order exchange is not allowed in magnetic systems due to its lack of time-reversal symmetry. Conversely, in classical fluids and chemisorbed overlayers, which can be modelled by magnetic Ising models via a lattice gas representation, three-particle interactions are known to contribute significantly to the thermodynamics (Rushbrooke 1968, Weinberg 1983). A well-known example of a magnetic system which is strongly influenced by fourth-order spin interactions is solid ^3He (Roger et al. 1983). An entirely new dimension to the formal study of fourth-order exchange has been added recently by the formulation of Ising lattice gauge theories involving four-spin interactions as representing the action associated with the Wilson loops of the lattice (see e.g. Savit 1980 for a review). Exploration of the phase structure of these lattice gauge theories has been important for understanding seminal aspects of elementary particle physics.

3.2.2 Ising models with multi-spin interactions

In the following, we shall be concerned with Ising models as described by the Hamiltonian

$$H = -J_2 \sum_{\{i,j\}} \sigma_i \sigma_j - J_3 \sum_{\{i,j,k\}} \sigma_i \sigma_j \sigma_k - J_4 \sum_{\{i,j,k,l\}} \sigma_i \sigma_j \sigma_k \sigma_l, \qquad (3.2.1)$$

where $\sigma_i = \pm 1$ and the three summations are extended over certain m-spin clusters ($m = 2, 3, 4$) on a lattice. Two- as well as three-dimensional lattices of different structures will be considered. We shall here think of Eq. (3.2.1) as a pure-model Hamiltonian leaving out of consideration the specific material it may describe.

Within the theory of phase transitions and critical phenomena, the tremendous interest in models with multi-spin interactions demonstrated over the past decade was sparked off by Baxter's exact solution of the eight-vertex model (Baxter 1971). Subsequent work (Kadanoff and Wegner 1971, Wu 1971) showed that the eight-vertex model is equivalent to two two-dimensional Ising models with nearest-neighbor exchange (J_2) coupled with one another by a four-spin coupling (J_4). A remarkable property of the Baxter model is that it exhibits non-universal critical behavior in the sense that the critical exponents vary continuously with the ratio J_4/J_2. The universality hypothesis for systems undergoing continuous transitions predicts that the critical exponents are independent of variations of linear parameters in the Hamiltonian, except at isolated points in the parameter space. Such isolated points may be characterized by changes in the nature of the phase transition, by changes in the symmetry of the order parameter, or by changes in the effective spatial dimension. Systems not covered by this prediction are those fulfilling the Kadanoff-Wegner criterion for non-universality (Kadanoff and Wegner 1971). This criterion states that a continuous variation of the critical exponents is allowed if the system Hamiltonian contains an operator which scales as r^{-d} under length-scale transformations. The Baxter model is a unique case of a model which contains such an operator. Certain square Ising lattices with pair interactions are also expected to display non-universal critical behavior (Jüngling 1976, Nightingale 1977, Domany and Riedel 1978, Landau 1980).

No three-dimensional system is definitely known to display non-universal critical behavior. However, in a series of papers on series analysis, Griffiths and Wood (1973, 1974, Wood and Griffiths 1974) have suggested that three-dimensional Ising models on cubic lattices with pair and multi-spin (three-spin and four-spin) interactions are possible candidates, despite the fact that these models do not fulfil the Kadanoff-Wegner criterion for non-universality. The resolution of this puzzle has been provided by computer simulations (Mouritsen et al. 1981a,b, 1983a, Frank and Mouritsen 1983). The computer simulations show that the continuous variation of the exponents found by Griffiths and Wood is an artifact of the series analysis and that the models display tricritical behavior. Along the line of continuous transitions in the phase diagram, universality is obeyed. The universality and tricritical behavior of cubic Ising lattices with two- and four-spin interactions is the subject of Sec. 3.2.4 below.

3.2.3 First-order phase transitions of Ising models with pure multi-spin interactions

Except for lattice gauge theories, models with pure multi-spin interactions are of limited physical relevance. However, from a theoretical standpoint such models are of considerable interest and constitute a challenge to theory on several accounts. Important questions to be answered are: i) what is the magnetic ground state and its degeneracy, ii) can pure multi-spin interactions alone support an ordered phase at non-zero temperatures, and iii) what is the nature of a possible phase transition.

Several *two-dimensional* Ising models with pure multi-spin interactions are known not to give rise to phase transitions. Examples are the quadratic Ising lattice with four-spin interactions around each basic square (i.e. the $J_2 \rightarrow 0$ limit of the Baxter model) and the honeycomb lattice with either four-spin interactions around each vertex (Wu 1972) or six-spin interactions around each hexagon (Griffiths and Wood 1973). In contrast, the triangular Ising lattice with three-spin interactions around each basic triangle (known as the Baxter-Wu model which can be solved exactly) has an ordered phase and exhibits critical behavior different from that of the standard two-dimensional Ising model (Baxter and Wu 1973).

In the remainder of this section, we shall focus upon *three-dimensional* Ising models on cubic lattices with four-spin interactions. The models are defined by Eq. (3.2.1) with $J_2 = J_3 = 0$. The basic interacting four-spin clusters $\{i, j, k, l\}$ are given in Fig. 3.2.1. For each lattice, these basic clusters represent the simplest equivalent and spatially the most confined four-spin clusters (quartets) which can be embedded in the given lattice. An important difference between the lattices is that the basic quartets for the *fcc* lattice involve nearest-neighbor bonds only, whereas the quartets of the *sc* and *bcc* lattices involve nearest- as well as next-nearest neighbor bonds. This entails that the ground states of the *sc* and *bcc* lattices are determined uniquely by the ground state configuration of a single basic quartet. For the *fcc* lattice, this is not the case and a tremendous ground state degeneracy arises. In the »ferromagnetic« case $J_4 > 0$, which we focus on here, this means that a ground state for the *sc* and *bcc* lattices consists of quartets with sign combinations $(+ + + +)$ or $(+ + - -)$ exclusively, in contrast to the *fcc* lattice where a ground state may involve mixtures of the two.

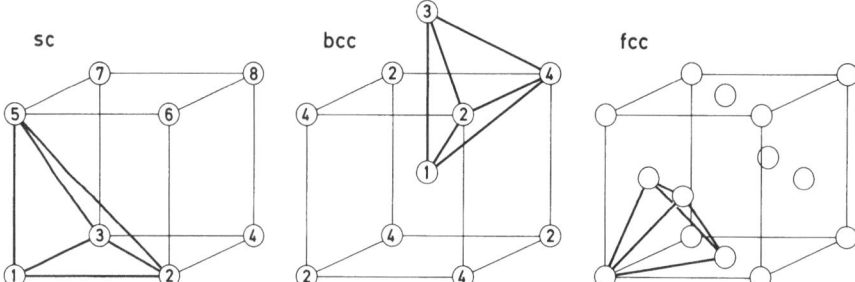

Fig. 3.2.1. Definition of the basic four-spin clusters (quartets) for the three cubic lattices. The basic quartets specify the four-spin interaction in Eq. (3.2.1) and are indicated by heavy solid lines. The numbers on the vertices indicate the sublattice labeling used for the ground state analysis.

An exact derivation of the ground states for the three cubic models has been given by Mouritsen et al. (1983a). The derivation is fairly straightforward for the *sc* and *bcc* lattices since it suffices to consider a single quartet. It turns out that the *sc* ground state is described by an $n = 8$-component order parameter and that the *bcc* lattice has an $n = 4$-fold degenerate ground state. In contrast, the ground state analysis of the *fcc* model is highly non-trivial and leads to a ground state degeneracy

$$W = 2^{3N^{1/3} - 2} \tag{3.2.2}$$

where N is the number of lattice points. The dimensionality of the order parameter can be shown to be $n = 3N^{1/3} - 2$ which tends towards infinity in the thermodynamic limit. Nevertheless, the ground state degeneracy is weak in the sense that there is no macroscopic entropy at zero temperature (i.e. zero residual entropy per spin)

$$S(T = 0)/N = N^{-1} k_B \ln W$$
$$= N^{-1} k_B (3N^{\frac{1}{3}} - 2) \ln 2 \to 0, \quad N \to \infty. \tag{3.2.3}$$

The *fcc* model has an unusual symmetry which is neither global nor local. From the construction of the ground states (Mouritsen et al. 1983a), it follows explicitly that one needs to fix spins of the order of $N^{\frac{1}{3}}$ in the lattice to determine uniquely a ground state (»to fix the gauge«). Thus, the model does not have a local symmetry as in gauge theories. This unusual choice of gauge has led Alcaraz et al. (1983) to suggest that the Z_M-symmetric generalizations of the *fcc* model (Ising: $M = 2$) constitute a three-dimensional theory which may naturally interpolate between two-dimensional globally-symmetric Z_M spin models and the four-dimensional locally-symmetric Z_M gauge theories.

In view of the current interest in systems with highly degenerate ground states, in particular fully-frustrated models (see e.g. Rácz and Vicek 1983 for a list of references), a comparison is in order between the *fcc* model with four-spin interactions and other models with infinitely degenerate ground states. These models are of interest since they may serve as models for spin glasses with non-random interactions (see e.g. Villain 1977). It should be emphasized, though, that our *fcc* model is not a frustrated model: every basic quartet in each ground state is in its lowest state of energy, $-J_4$. To the best of our knowledge, this model is the only non-frustrated model with infinitely many ground states which so far has had its ground state problem worked out systematically. Several *two-dimensional* fully-frustrated models have been described, both with zero and non-zero residual entropy (Derrida et al. 1980, Forgacs 1980, Berker and Kadanoff 1980, Chui et al. 1982, Rácz and Vicek 1983). Some of these models have conventional long-range order at finite temperatures and some have not. The classical example is the triangular antiferromagnetic Ising lattice solved exactly by Wannier (1950). This model has non-zero residual entropy and is disordered at all finite temperatures. The general principles relating the nature of the infinite ground state degeneracy to the existence and nature of phase transitions have not yet been discovered. Two *three-dimensional* fully-frustrated Ising models have had their ground state properties systematically worked out. One of these is the *fcc* Ising antiferromagnet with nearest-neighbor interactions (Danielian 1961, 1964, Alexander and Pincus 1980) which is the classical model of *CuAu*-type alloys (Binder 1983). The other one is the fully-frustrated *sc* Ising model with competing nearest-neighbor ferromagnetic and antiferromagnetic interactions (Chui et al. 1982). The ground state degeneracy for these two models are respectively of the order of $W \sim 2^{N^{\frac{1}{3}}}$ and $W \sim 2^{N^{\frac{2}{3}}/4}$, and in analogy with our *fcc* quartet model neither of these two models has a macroscopic entropy at zero temperatures. As we shall describe below, there is mounting evidence that these two frustrated models as well as the *fcc* model with four-spin interactions exhibit conventional finite-temperature long-range order in spite of the infinite degeneracy of the order parameter.

The question whether a finite-temperature transition exists in the three cubic Ising lattices with pure four-spin interactions has been approached via low-temperature series analysis for the *fcc* and *bcc* lattices (Griffiths and Wood 1973, 1974, Wood and Griffiths 1974) and for all three cubic lattices by Monte Carlo simulations (Mouritsen et al. 1981a, 1983a, Liebman 1982, Alcaraz et al. 1982, 1983). Of these two approaches, only the Monte Carlo calculations are capable of establishing directly the existence of finite-temperature transitions. In the low-temperature series analysis work, only the consequences of assuming the existence of a continuous transition can be explored.

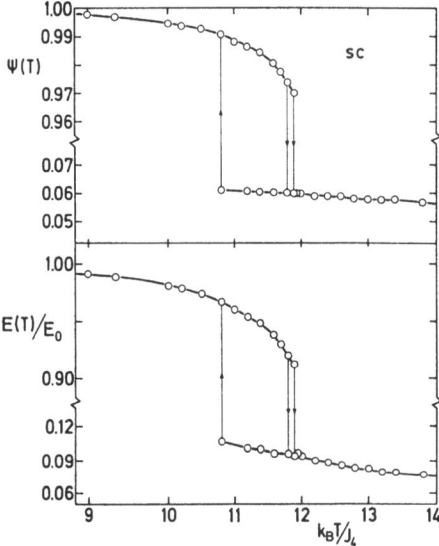

Fig. 3.2.2 Order parameter $\psi(T)$ and normalized internal energy $E(T)/E_o$ vs temperature for the *sc* Ising lattice with pure four-spin interactions. The data points are obtained from Monte Carlo calculations on a lattice with $N = 216$ spins. Vertical arrows indicate directions of transitions observed for increasing and decreasing temperature series.

The Monte Carlo results for the *sc* and *bcc* lattices will be presented first. The internal energy and the various components of the order parameter have been calculated for increasing as well as decreasing series of temperatures. The result for a *sc* lattice with $N = 216$ spins are presented in Fig. 3.2.2. This figure demonstrates that the model has a finite-temperature phase transition. Two branches of the curves are observed corresponding to the increasing and decreasing temperature series. Close to the termini of the branches, pronounced discontinuities are encountered for both functions. Together with the occurrence of metastable states and coexistence of phases, this unequivocally demonstrates that the transition is of first order. Within our observation time (< 14000 Monte Carlo steps per site), the order on the low-temperature branch resides in one of the eight components, ψ_i, and the one under consideration is then termed *the* order parameter $\psi(T)$. The values of the remaining components, ψ_j, $j \neq i$, are the same and small, but finite, due to the finite size of the lattice. On the high-temperature branch, the finite-size order is distributed equally among the eight components. When the system undergoes the transition from the disordered phase to the ordered phase, it is equally likely to enter any one of the eight degenerate ordered states and the corresponding finite-temperature internal energy is found to be independent of whichever order parameter component becomes dominant. This shows that the eight-fold degeneracy of the *sc* lattice is retained for all temperatures in the ordered phase.

The hysteresis loops in Fig. 3.2.2 for the *sc* lattice cover an extended temperature range and it is therefore not possible directly from these static data to estimate accurately the equilibrium transition temperature. Consequently, we have performed a calculation of the lifetime, τ, of the metastable states in the transition region. Defining τ as the number of Monte Carlo steps per site to be performed before a metastable state undergoes the transition to the stable state in course of the characteristic two-step relaxation process (cf. Sec. 2.2.9), we have obtained the curve $\tau^{-1}(T)$ displayed in Fig. 3.2.3. A pronounced asymmetric relaxation behavior is observed and the relaxation times on the disordered metastable branch are found to be extremely long. The equilibrium transition temperature is the temperature where τ^{-1} attains its minimum. Alternatively, the equilibrium transition temperature can be determined from the free energy function, $F(T)$, which in turn may be derived from the internal energy using the relationship $T dF = dE$. For the two phases, we then have the formulas (cf. Eq. (2.2.28))

$$F(T) = E_o + T \int_0^{1/T} [E(T) - E_o] \, d(1/T), \quad T < T_c \tag{3.2.4}$$

and

$$F(T) = -T S_\infty + T \int_\infty^{1/T} E(T) \, d(1/T), \quad T > T_c, \tag{3.2.5}$$

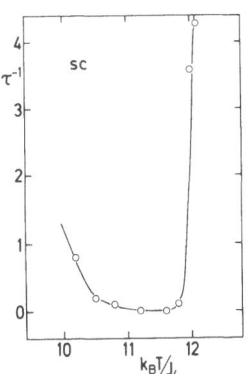

Fig. 3.2.3. Inverse relaxation time τ^{-1} in arbitrary units *vs* temperature in the transition region for the *sc* Ising lattice with pure four-spin interactions. The data are obtained from Monte Carlo calculations on a system with $N = 216$ spins.

where E_o is the ground state energy and $S_\infty = Nk_B \ln 2$ is the infinite-temperature entropy. In performing the integration in Eqs. (3.2.4) and (3.2.5), we use that for $k_B T/J_4 \gtrsim 12$ the Monte Carlo data for $E(T)$ coalesce with the first term in the high-temperature series expansion, $E(T) = -2J_4/k_B T + \dots$, and that deviations from E_o are negligible for $k_B T/J_4 \lesssim 2$. Results from the numerical integration of the Monte Carlo data are given in Fig. 3.2.4. The equilibrium transition temperature is obtained from the intersection point of the two free energy branches. The result from this procedure is consistent with the life-time measurements in Fig. 3.2.3.

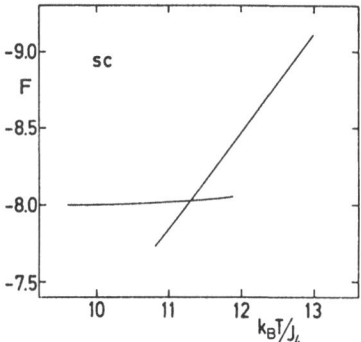

Fig. 3.2.4. Free energy $F(T)$ (in units of NJ_4) vs temperature for the *sc* Ising lattice with pure four-spin interactions. $F(T)$ is obtained from a numerical integration of the internal energy in Fig. 3.2.2.

The Monte Carlo results for the *bcc* lattice are similar to those reported for the *sc* lattice. There is clear evidence of a finite-temperature first-order phase transition. The ground state degeneracy is found to remain operative throughout the ordered phase.

Finally, we turn to the *fcc* Ising lattice with pure four-spin interactions. This model is of particular interest since it has an infinitely degenerate ground state. The model was conjectured already some time ago to be self-dual (Wood 1972) and proof of this self-duality has recently been given (Liebman 1981, Pearce and Baxter 1981). The self-dual property is expressed in terms of the self-duality relation

$$\exp(-2J_4/k_B T) = \tanh(J_4/k_B T^*) \tag{3.2.6}$$

which relates possible high- to possible low-temperature singularities of the partition function (Wannier 1945). Equation (3.2.6) implies that the *fcc* model is self-dual with respect to the four-spin interactions in the same way as the square Ising lattice is self-dual with respect to nearest-neighbor pair interactions (see e.g. Syozi 1972). Hence, if the model has finite-temperature phase transitions it has either two transition points related by Eq. (3.2.6) or a single transition at the Onsager point ($T = T^* = T_c^O$ in Eq. (3.2.6))

$$k_B T_c^O/J_4 = -2/\ln(\sqrt{2} - 1) \simeq 2.27. \tag{3.2.7}$$

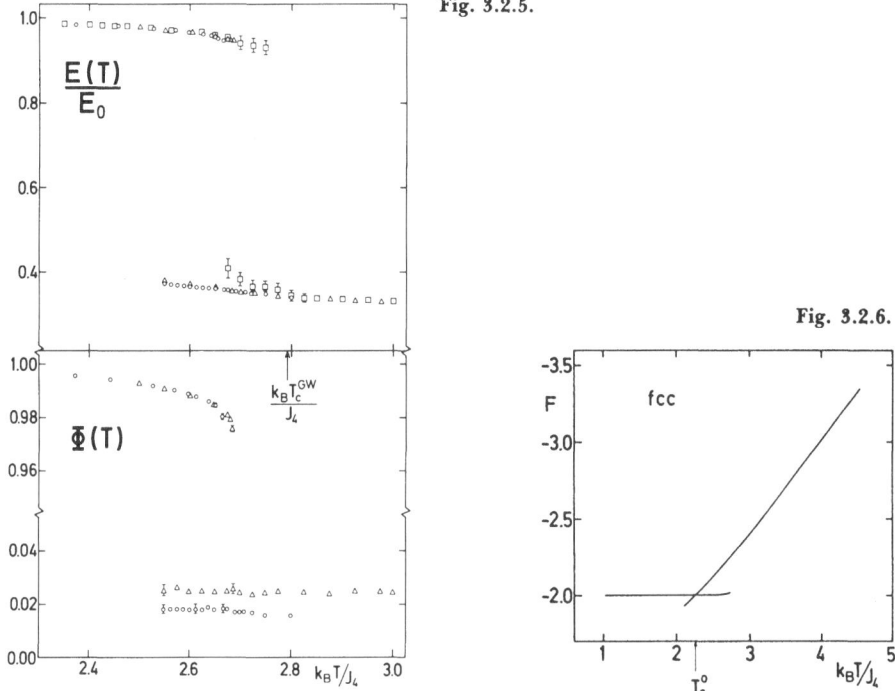

Fig. 3.2.5. Normalized internal energy $E(T)/E_o$ and ferromagnetic order parameter $\Phi(T)$ vs temperature for the *fcc* Ising lattice with pure four-spin interactions. The data are obtained from Monte Carlo calculations on lattices with N spins. $N = 128$ (squares), $N = 1024$ (triangles), and $N = 2000$ (circles). T_c^{GW} is the transition temperature from the series analysis by Griffiths and Wood (1973).

Fig. 3.2.6. Free energy $F(T)$ (in units of NJ_4) vs temperature for the *fcc* Ising lattice with pure four-spin interactions. T_c^0 denotes the Onsager point, Eq. (3.2.7).

From their series analysis, Griffiths and Wood (1973) determined the transition temperature to be $k_B T_c^{GW}/J_4 = 2.79 \pm 0.01$. They concluded that the model has two transition temperatures. The earliest Monte Carlo calculations on the *fcc* model demonstrate that this conclusion is wrong (Mouritsen et al. 1981a). In Figs. 3.2.5 and 3.2.6 are shown the Monte Carlo results for the internal energy, the order parameter, and the free energy. Obviously, the model undergoes a single and very pronounced first-order phase transition. The equilibrium transition temperature is found from the free energy function in Fig. 3.2.6 to be $k_B T_c^{MC}/J_4 = 2.27 \pm 0.02$, i.e. precisely at the Onsager self-dual point, Eq. (3.2.7) (Mouritsen et al. 1983a). Similar conclusions have been reached from other independent Monte Carlo studies (Liebman 1982, Alcaraz et al. 1982, 1983).

Our findings of first-order phase transitions in all three cubic Ising lattices with pure four-spin interactions are in accordance with the prediction of simple Landau (or mean-field) theory. This is seen by equalizing all n ordering fields (denoted ϕ) and writing the Landau free energy as

$$4F = -n_q J_4 \phi^4 + 2k_B T[(1 + \phi)\ln(1 + \phi) + (1 - \phi)\ln(1 - \phi)]$$
$$= 2k_B T\phi^2 + (\tfrac{1}{3}k_B T - n_q J_4)\phi^4 + \dots, \tag{3.2.8}$$

where n_q is the number of quartets a given spin participates in ($n_q = 32, 24$, and 8 for the *sc*, *bcc*, and *fcc* lattices). This free energy leads to a first-order transition since the second-order term is positive. A renormalization group analysis in $4 - \epsilon$ dimensions of the completely symmetrized Landau-Ginsburg-Wilson Hamiltonians, which can be constructed for the *sc* and *bcc* lattices (Mouritsen et al. 1983a), is not expected to change this prediction.

A finite-temperature first-order phase transition has also been observed in the fully-frustrated *fcc* Ising antiferromagnet by Monte Carlo calculations (Phani et al. 1979, 1980a). Mean-field theory predicts a continuous transition for this model. Chui et al. (1982) studied the fully-frustrated *sc* Ising model and argued, on the basis of the Bethe-Peierls approximation, that this model has a finite-temperature continuous transition. This is in accordance with the results of a Monte Carlo study (Bhanot and Creutz 1980). However, Chui et al. asserted that fluctuations should drive the transition first order and claimed that the Monte Carlo result could be misleading due to the degeneracy of the ground state. This, then, prompts us to address the problems which may arise in computer simulations of models with infinitely many degenerate ground states.

Obviously, a finite lattice can only accomodate a finite number of ordered states. One could therefore speculate that computer simulations on finite lattices may give a distorted picture of the true thermodynamic limit. Calculations on a series of different lattice sizes for the two different *fcc* models considered here indicate that the results do not depend significantly on the lattice size. A similar conclusion is drawn from Monte Carlo studies of two- and three-dimensional $q \geq 3$-state antiferromagnetic Potts models (Rácz and Vicek 1983, Banavar et al. 1980, Grest and Banavar 1981) for which the ground state degeneracy is even stronger. Furthermore, the fact that the exact self-dual property is reproduced by the Monte Carlo simulations on the finite *fcc* quartet model strengthens the evidence that finite lattices reflect the true phase structure of the model. In summary, we find no evidence that finite-lattice Monte Carlo simulations should be invalid for systems with infinite ground state degeneracy. Therefore, Chui et al.'s (1982) interpretation of the Monte Carlo work on the fully-frustrated *sc* model seems barely possible to us.

For the *fcc* model with four-spin interactions, the number of degenerate ground states is given by Eq. (3.2.2). This is a large number even for a small lattice. Consequently, we have only calculated the ferromagnetic component which is the one plotted in Fig. 3.2.5. In the calculations, the lattice is initiated in the ferromagnetic state at low temperatures. Whenever the system enters the disordered state or any other ordered state, information is not obtained about the actual long-range order but rather about the noise in a non-ordering component (the lower branch of $\Phi(T)$ in Fig. 3.2.5). Similar comments apply to the Monte Carlo study of the fully-frustrated *fcc* antiferromagnet (Phani et al. 1979). The high degeneracy of the ordered state makes it extremely difficult in the Monte Carlo calculations to bring the system into an ordered phase in decreasing temperature series. Thus extremely wide hysteresis loops are encountered. This situation forced the authors of the earliest computer study of this model (Mouritsen et al. 1981a) to erroneously estimate the transition temperature from the terminus of the upper branches in Fig. 3.2.5. Only by a free energy calculation (Mouritsen et al. 1983a, Liebman 1982) or by using a mixed phase method (cf. Sec. 2.2.9) (Alcaraz et al. 1982) to shorten the life times of the metastable states, can a reliable determination of the equilibrium transition temperature be provided.

In closing the section on models with pure multi-spin interactions, we shall briefly describe the results of a Monte Carlo study of a *fcc* Ising lattice with pure three-spin interactions ($J_2 = J_4 = 0$) in Eq. (3.2.1) (Mouritsen et al. 1981a). The three-spin clusters $\{i, j, k\}$ of Eq. (3.2.1) comprise all elementary triangles of the *fcc* lattice. Again, the Monte Carlo calculations predict unambiguously the existence of a single finite-temperature first-order phase transition. This finding is in accordance with a simple mean-field calculation. Also for this model a low-temperature series analysis has been reported (Griffiths and Wood 1973) to yield critical behavior characterizing a new universality class. The Monte Carlo study shows that the basic assumption underlying the series analysis in terms of critical behavior is not valid.

To conclude this section, it can on the one hand be said that by means of computer simulations we have shown that higher-order exchange in cubic Ising models leads to first-order transitions. The same statement is believed to hold for Heisenberg models (Adler and Oitmaa 1979). On the other hand, one should be careful about explaining experimental observations of first-order phase transitions in magnetic materials, in which higher-order exchange is known to be present, as direct consequences of the higher-order exchange. Firstly, more fundamental explanations may be appropriate. An example is the first-order magnetic phase transition detected in solid ^3He (Osheroff et al. 1980). This can be explained by the lack of stable fixed points in the Landau-Ginzburg-Wilson Hamiltonian which can be constructed from the $n = 12$-component order parameter pertinent for the type-IV *bcc* antiferromagnetic structure (Bak and Rasmussen 1981) (cf. Sec. 4.2.1). According to the renormalization group argument, the transition should be of first order irrespective of the nature of the microscopic interactions. Secondly, the simultaneous presence of pair interactions can change the nature of the phase transition into a continuous transition. This change will be the subject of the following section.

3.2.4 Universality and tricritical behavior of Ising models with two- and four-spin interactions: Pair interactions as a symmetry-breaking field

With very few exceptions, the conventional pair interaction approximation is an extremely good approximation for describing the properties of magnetic materials and alloys. Nevertheless, higher-order interactions are always present to some degree. The results presented in the preceeding section show that the phase behavior governed by pure higher-order interactions is often qualitatively different from that produced by pure pair interactions. The question naturally arises to which extent the presence of multi-spin interactions influences the phase behavior. Ising models as defined by Eq. (3.2.1) are ideal and simple models to help answer this question.

The two-dimensional Baxter model is an exceptional case of a model for which the simultaneous presence of two- and four-spin interactions induces non-universal critical behavior. Conversely, any finite amount of pair interactions added to the three-spin interactions in the Baxter-Wu model changes the critical behavior to that characteristic of pure pair interactions (Imbro and Hemmer 1976). Here, we address questions related to the phase behavior of *three-dimensional* cubic Ising lattices with mixtures of pair and four-spin interactions ($J_3 = 0$ in Eq. (3.2.1)).

We shall think of the ferromagnetic pair interactions in Eq. (3.2.1) as an effective field which breaks the symmetry of the order parameter of the pure four-spin interaction models. Two types of symmetry-breaking fields will be considered: (i) isotropic nearest-neighbor ferromagnetic pair interactions in all three cubic lattices, and (ii) nearest-neighbor ferromagnetic pair interactions within one or three of the four *fcc* sublattices which make up the *bcc* lattice. These two fields break the symmetry in two different ways. In the case of (i), the $n = 1$ ferromagnetic representation is singled out and stabilized. In the case of (ii), the $n = 4$ degeneracy of the *bcc* model is split into an $n = 1$ and an $n = 3$ representation. Case (i) will be considered first.

Mouritsen et al. (1981b) and Frank and Mouritsen (1983) have studied, by means of computer simulations and analytical calculations, the phase transition in all three cubic Ising lattices with mixtures of isotropic pair and four-spin interactions. A wide range of lattice sizes, ranging form $N = 216$ to 16000 spins, were studied in order to reveal the nature of the transitions as a function of J_4/J_2. As an example, detailed results for the *fcc* lattice will be presented (Mouritsen et al. 1981b). In Fig. 3.2.7 is shown the ferromagnetic order parameter, $\Phi(T)$, as a function of

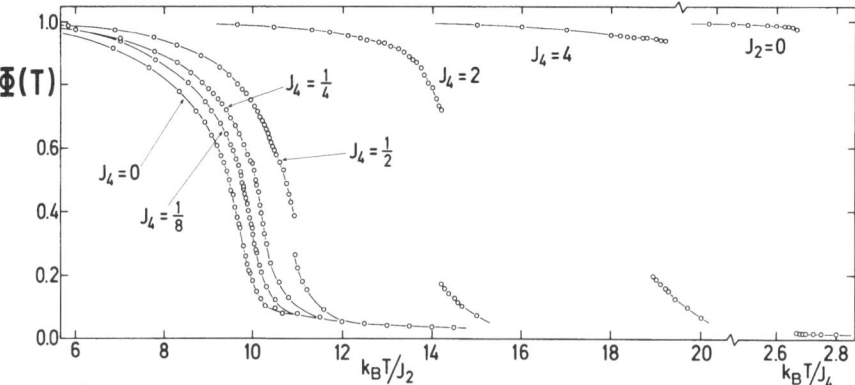

Fig. 3.2.7. Ferromagnetic order parameter, $\Phi(T)$, *vs* temperature for the *fcc* Ising model with pair (J_2) and four-spin (J_4) interactions. The data are obtained from Monte Carlo calculations on systems with $N = 2000$ spins, except for $J_4 = 4$ where the system contains $N = 432$ spins.

temperature for various values of the coupling ratio J_4/J_2. These results (and the corresponding results for the internal energy) demonstrate that the *first-order phase transition* found in the pure four-spin interaction limit survives for $J_4/J_2 \gtrsim \frac{1}{2}$. For decreasing values of J_4/J_2 in the first-order regime, the discontinuity in the order parameter decreases, and eventually for $J_4/J_2 \lesssim \frac{1}{4}$ it disappears. For $J_4/J_2 \lesssim \frac{1}{4}$, we conclude, by using the criteria given in Sec. 2.2.9, that the phase transition is *continuous*. Thus, the model possesses a *tricritical point* somewhere in the region $\frac{1}{4} < J_4/J_2 < \frac{1}{2}$. A detailed graph of the internal energy close to the tricritical point is given in Fig. 3.2.8. This figure shows clearly that all characteristics of a first-order transition are absent for $J_4/J_2 = \frac{1}{4}$.

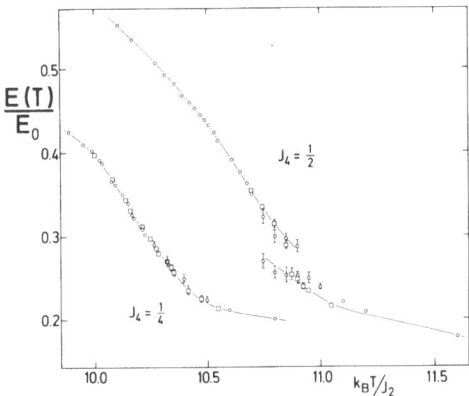

Fig. 3.2.8. Normalized internal energy, $E(T)/E_0$, *vs* temperature in the tricritical region of the *fcc* Ising model with pair (J_2) and four-spin interactions (J_4). The data are obtained from Monte Carlo calculations on systems with N spins. Circles indicate $N = 2000$, squares indicate $N = 11644$ ($J_4/J_2 = \frac{1}{4}$) and $N = 16000$ ($J_4/J_2 = \frac{1}{2}$).

The critical behavior of the order parameter in the region of continuous transitions is investigated in Fig. 3.2.9 which is a log-log plot of $\Phi(t)$ *vs* reduced temperature, $t = (T_c - T)/T_c$. The critical temperatures needed for this analysis are those derived from the series analysis of the model (Griffiths and Wood 1974, Fisher and Burford 1967). In the regime of continuous transitions, the estimates of the critical temperatures from series analysis are considered to be more accurate than the Monte Carlo results (whereas the opposite is true in the first-order regime). From a study of the finite-size effects, we conclude that the data in Fig. 3.2.9 describe rather accurately the thermodynamic limit for $t \gtrsim 0.008$. The figure shows that the data for $J_4 = 0$ satisfy the expected asymptotic form

$$\Phi(t) \simeq Bt^\beta \tag{3.2.9}$$

with critical parameters $\beta = 0.310 \pm 0.015$ and $B = 1.45 \pm 0.05$. These parameters are consistent with the respective series estimates, $\beta = 0.312 \pm 0.005$ and $B = 1.487 \pm 0.002$ (Domb 1974), for the *fcc* Ising model with pure pair interactions.

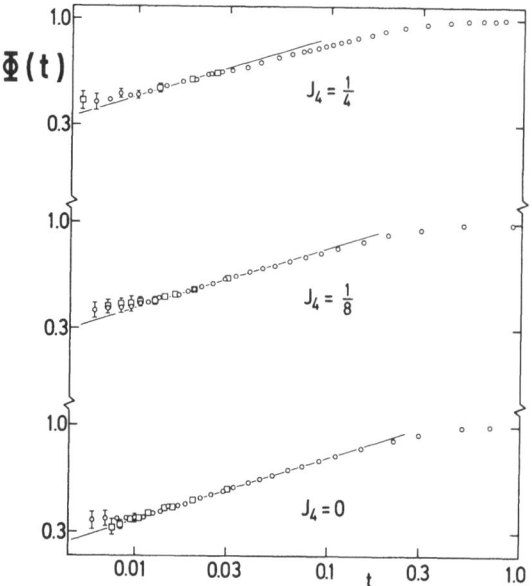

Fig. 3.2.9. Log-log plot of the ferromagnetic order parameter $\Phi(t)$ *vs* reduced temperature, $t = (T_c - T)/T_c$, for the *fcc* Ising model with pair (J_2) and four-spin (J_4) interactions. The data are obtained from Monte Carlo calculations on systems with $N = 2000$ spins (circles) and $N = 11664$ spins (squares). The solid lines represent simple power laws, $\Phi(t) \sim t^\beta$, with $\beta = 0.31$.

For increasing values of J_4/J_2, Fig. 3.2.9 shows that the asymptotic slope remains unchanged, but the amplitude B increases and the critical region described by Eq. (3.2.9) shrinks. The shrinking is caused by crossover to the tricritical point. The data are considered to be too far from the tricritical point to sustain a detailed analysis of $\Phi(t)$ in terms of multiplicative logarithmic corrections (Wegner and Riedel 1973; for a Monte Carlo study of logarithmic corrections to tricritical behavior, see Landau 1976c).

Monte Carlo calculations on the *sc* and *bcc* Ising lattices with mixtures of pair and four-spin interaction suggest that also these lattices exhibit tricritical behavior (Frank and Mouritsen 1983). In fact, the phase behavior of all three lattices is remarkably similar even quantitatively. This is demonstrated in Fig. 3.2.10 which shows a scaled plot of the composite phase diagram encompassing transitions for all three lattices. The diagram is presented in terms of a set of mean-field-like scaled variables

$$\tilde{J} = y/(1 + y) \tag{3.2.10}$$

$$\tilde{T}_c = J_2(\text{pair})T_c/[J_2 T_c(\text{pair})(1 + y)] \tag{3.2.11}$$

$$y = n_q J_4/n_p J_2. \tag{3.2.12}$$

n_q and n_p is the number of quartets and pairs which a given spin participates in. The phase boundary for all three lattices may be fitted by a common curve. Moreover, this curve is approximately linear. The linearity and the approximate

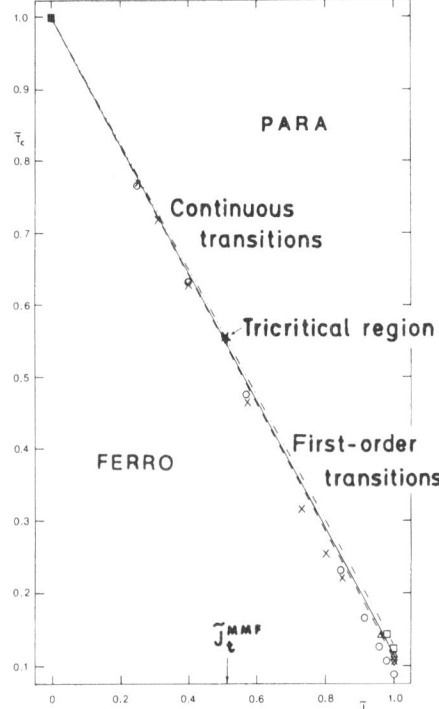

Fig. 3.2.10. Scaled phase diagram $(\bar{J}, \bar{T}_c)$ for Ising models on cubic lattices with mixtures of pair and four-spin interactions. The diagram gives the phase boundary between the ferromagnetic and paramagnetic phases and contains a region of continuous transitions as well as a region of first-order transitions, separated by a tricritical point near the asterix. The scaled parameters are defined in Eqs. (3.2.10) - (3.2.12). Circles, triangles, and squares are Monte Carlo results for the *fcc*, *bcc*, and *sc* lattices, respectively. Series results are given for the *fcc* (crosses) and *bcc* (inverse triangles) lattices. The lines are the theoretical predictions by Frank and Mouritsen (1983) for the *fcc* (dashed), *bcc* (solid), and *sc* (dot-dashed) lattices. $\bar{J}_t^{MMF}$ is the tricritical value of $\bar{J}$ as predicted by a modified mean-field theory (Mouritsen et al. 1981b).

lattice-lattice scaling of the phase boundary have been explained within a recent analytical theory of the critical correlation functions (Frank and Mouritsen 1983). The tricritical behavior of Ising models with mixtures of pair and four-spin interactions is also quite generally predicted by renormalization group studies (Aharony 1974, Gitterman and Mikulinsky 1977). Furthermore, it is in qualitative accordance with simple Landau theory. This is most easily seen from an appropriate modification of the free energy in Eq. (3.2.8) according to

$$4F = 2(k_B T - n_p J_2)\phi^2 + (\tfrac{1}{3}k_B T - n_q J_4)\phi^4 + \ldots . \qquad (3.2.13)$$

A tricritical point is predicted within Landau theory when $\partial^2 F/\partial\phi^2 = \partial^4 F/\partial\phi^4 = 0$. This leads to a tricritical coupling ratio $(J_4/J_2)_t = n_p/3n_q$ and to transition temperatures which are independent of J_4/J_2. These predictions are in quantitative disagreement with the Monte Carlo results. A more satisfactory agreement with the Monte Carlo results can be obtained from a modified mean-field theory which takes account of third-order correlations (Mouritsen et al. 1981b).

By predicting tricritical behavior of cubic Ising models with mixtures of pair and four-spin interactions, the Monte Carlo studies resolve a puzzle which has persisted for quite some time in the literature of phase transitions. This puzzle was brought about by the evidence reported from series analysis (Griffiths and Wood 1974) in favor of non-universal critical behavior for these models. The non-universal behavior is simply an artifact of the series analysis applied to first-order transitions assuming power-law singularities. More importantly, the Monte Carlo calculations provide strong evidence for the universality hypothesis: the critical behavior (i.e. the exponents) is independent of the coupling ratio in the regime of continuous transitions (cf. Fig. 3.2.9).

Next, we consider the case of the *bcc* lattice with four-spin interactions and a symmetry-breaking field, g, which splits the $n = 4$ representation into an $n = 1$ and an $n = 3$ representation. The field corresponds to a magnetic field or a uniaxial stress which only couples to even orders in the order parameter. As emphasized by Kerszberg and Mukamel (1981), it is of interest to study, for this symmetry, the influence of the field g on the properties of the phase transition. Thus, a Monte Carlo study has been conducted to map out the phase diagram in the (T, g)-plane (Mouritsen et al. 1983a). In the microscopic Hamiltonian, g is simulated by suitable pair interactions defined on subsets of the four inter-penetrating *fcc* sublattices given in Fig. 3.2.1. The four-spin term couples these four sublattices. Here we single out one or three sublattices by introducing ferromagnetic *fcc* nearest-neighbor (*bcc* third-nearest neighbor) interactions in the sublattices. The effective Hamiltonian then takes the form

$$H = -J_2^+ \overset{(1)}{\underset{\{i,j\}}{\sum}} \sigma_i \sigma_j - J_2^- \overset{(2+3+4)}{\underset{\{i,j\}}{\sum}} \sigma_i \sigma_j - J_4 \underset{\{i,j,k,l\}}{\sum} \sigma_i \sigma_j \sigma_k \sigma_l,$$

$$J_2^+ > 0, \quad J_2^- > 0,$$

(3.2.14)

where the first sum is defined on sublattice 1 and the second sum on sublattices 2 – 4.

In terms of the sublattice order parameters ϕ_1 - ϕ_4, the corresponding Landau free energy functional may be written (to fourth order in ϕ)

$$4F = -n_q J_4 \phi_1 \phi_2 \phi_3 \phi_4 + 2(k_B T - n_p J_2^+)\phi_1^2$$

$$+ 2(k_B T - n_p J_2^-) \overset{4}{\underset{i=2}{\sum}} \phi_i^2 + \tfrac{1}{3} k_B T \overset{4}{\underset{i=1}{\sum}} \phi_i^4.$$

(3.2.15)

n_p is the number of nearest neighbor pairs per spin of the *fcc* lattice ($n_p = 6$). For convenience, we introduce the scaling fields r and g defined by

$$r - 3g = 2k_B T - 2n_p J_2^+$$
$$r + g = 2k_B T - 2n_p J_2^-,$$

(3.2.16)

i.e.

$$g = \tfrac{1}{2} n_p (J_2^+ - J_2^-)$$
$$r = 2k_B T - \tfrac{1}{2} n_p (J_2^+ + 3J_2^-).$$

(3.2.17)

r is the temperature-like variable. The mean-field phase diagram (r, g) resulting from the free energy in Eq. (3.2.15) has been calculated by Kerszberg and Mukamel (1981) for a similar $n = 4$ model and its qualitative characteristics are shown in Fig. 3.2.11. The diagram is rather complex including first-order and continuous transition lines, two critical end points, and a liquid-gas-like critical point. For low values of $|g|$, there is a single first-order transition from the four-fold degenerate phase, $[\phi_1, \phi_2, \phi_3, \phi_4]$, to the paramagnetic phase. For larger values of g, there is a four-state Potts-like first-order transition within the ordered phase to the one-component phase $[\phi_1]$. In this phase, the fluctuations of the fields ϕ_2, ϕ_3, and ϕ_4 are quenched and for high temperatures the system undergoes an $n = 1$ continuous Ising-like transition. For an intermediate range of negative values of g, there is also a first-order transition within the ordered phase. The corresponding phase line, which terminates in a liquid-gas-like critical point, is not associated with a change of symmetry. Since the order parameters ϕ_2, ϕ_3, and ϕ_4 induce a field $\sim \phi_2 \phi_3 \phi_4$ which couples linearly to ϕ_1, all four fields are non-zero on the high-temperature side of this transition. For still higher temperatures, there is a line of continuous phase transitions to the paramagnetic phase. This transition line, which belongs to the universality class of the $n = 3$ Heisenberg model with cubic anisotropy, persists for large negative values of g.

We have carried out Monte Carlo temperature scans for a series of values of g in order to map out the phase diagram. In Fig. 3.2.12 we give, in the cases of $g = 1.5$ and $g = 6$, the results for the internal energy and the two types of order parameters, ψ and ϕ_1. ψ is one of the four order parameter components

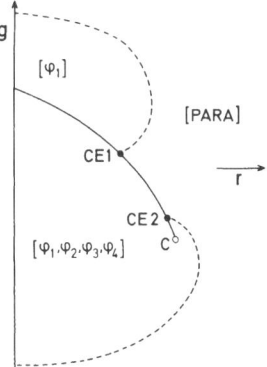

Fig. 3.2.11. Schematic phase diagram (r, g) for the model in Eq. (3.2.14) solved within the mean-field approximation. For explanation, see the main text.

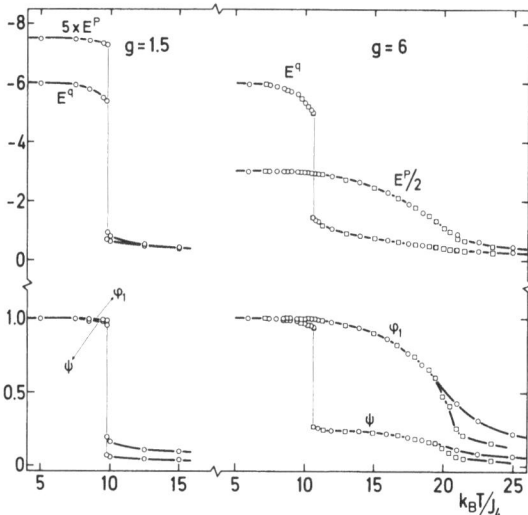

Fig. 3.2.12. Temperature dependence of internal energies and order parameters for a *bcc* Ising model with four-spin interactions and subjected to a positive symmetry-breaking field, g. E^p and E^q are the pair interaction and four-spin interaction energies. ψ is the order parameter governed by the four-spin interaction, and ϕ_1 is the order parameter governed by the pair interaction defined in sublattice 1 (cf. Fig. 3.2.1). Monte Carlo data are given for systems with $N = 432$ (circles) and $N = 2000$ (squares) spins. All energies are in units of NJ_4. g is simulated by pair interactions with $J_2^+ > 0$ and $J_2^- = 0$.

$$\psi_1 = \phi_1 + \phi_2 + \phi_3 + \phi_4$$
$$\psi_2 = \phi_1 + \phi_2 - \phi_3 - \phi_4$$
$$\psi_3 = \phi_1 - \phi_2 + \phi_3 - \phi_4 \qquad\qquad (3.2.18)$$
$$\psi_4 = \phi_1 - \phi_2 - \phi_3 + \phi_4.$$

The internal energy is split into the contributions from the pair interactions (E^p) and the four-spin interactions (E^q). For $g = 1.5$, the figure shows a pronounced discontinuity in all functions implying that the phase transition is of first order. A very narrow hysteresis loop is encountered. For $g = 6$, the behavior has changed drastically. Now we observe two consecutive transitions. The lower one is of first order signalled by discontinuities in E^q and ψ. This transition is described by an effective four-state Potts-type Hamiltonian, and our finding of a first-order transition accords with the theoretical predictions for three-dimensional Potts models with three or more components (Zia and Wallace 1975). The functions E^p and ϕ_1 pass smoothly through this first transition. At the upper transition, E^p and ϕ_1 change in a continuous manner, In going from $g = 1.5$ to $g = 6$, the discontinuities in E^q and ψ decrease. In Fig. 3.2.12, the data for $g = 6$ are given for two different lattice sizes, $N = 432$ and $N = 2000$. We note that the lower transition is not significantly affected by finite-size effects, which is expected for a first-order transition. However, the upper transition becomes significantly sharper but remains continuous when N is increased. ϕ_1 is more affected than E^p. Furthermore, we find that the specific heat peak increases in intensity and moves towards higher temperatures when the lattice size is increased. All this evidence is consistent with the upper transition being continuous. However, we do not have sufficient information to determine critical exponents and thereby identify the universality class. It should be noted that only one quarter of the system goes critical at this upper transition. The order parameter ψ is non-zero in the intermediate phase because ψ, via Eq. (3.2.18), involves ϕ_1. Therefore, also ψ decreases continuously at the upper transition, and the high-temperature tail of this order parameter, as well as that of ϕ_1, is due to conventional finite-size effects.

We now turn to negative values of the symmetry-breaking field g. For small negative g, we observe a single clear and pronounced first-order transition. In the case of larger negative g-values, $g = -7.5$ and $g = -12$, Fig. 3.2.13 displays the Monte Carlo results for E^p, E^q, and the order parameters ψ and $\phi = \frac{1}{3}(\phi_2 + \phi_3 + \phi_4)$. Data for $N = 432$ and $N = 2000$ are shown, and for the sake of clarity only results for the larger system are plotted for the order parameters close to and above the transition. For $g = -7.5$, simultaneous discontinuities in all four functions indicate the presence of a single first-order transition. However, for this value of g the finite-size effects are much more pronounced that for the values of g given in Fig. 3.2.12. These finite-size effects (not shown in the figure)

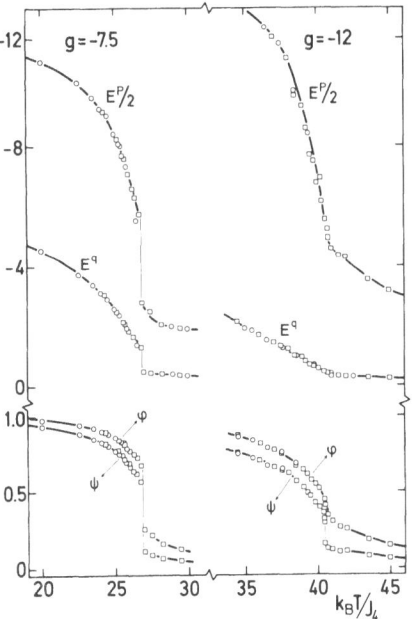

Fig. 3.2.13. Temperature dependence of internal energies and order parameters for a *bcc* Ising model with four-spin interactions and subjected to a negative symmetry-breaking field, g. ψ is the order parameter governed by the four-spin interactions, and ϕ is the order parameter governed by the pair interactions defined in sublattices 2 - 4 (cf. Fig. 3.2.1). The remaining symbols are explained in Fig. 3.2.12. g is simulated by pair interactions with $J_2^+ = 0$ and $J_2^- > 0$.

tend to diminish the discontinuities and lead to a partial smearing of the transition. Thus, we conclude that significant thermodynamic fluctuations have come to play a role and that the system may not be far from a critical point. For $g = -12$, Fig. 3.2.13 shows that the discontinuities in E^p, ϕ, and E^q have now disappered, but a small jump in ψ remains. The temperature and finite-size variation of E^p and ϕ show characteristics of a continuous transition, in contrast to that of ψ which definitely still undergoes a first-order transition, although E^q is now changing smoothly in the transition region. As always in numerical simulations, we cannot exclude that a seemingly continuous transition is actually a first-order one with a very small discontinuity which is veiled by the finite-size effects. However, we shall here assume that the transition in E^p and ϕ is continuous. More important is it that there appears to be a separation between the two transitions indicated by the fact that the inflection point of E^p and ϕ appears slightly above the jump in ψ. Thus, the data are consistent with two very close-lying transitions. This supports the mean-field phase diagram in Fig. 3.2.11. According to this diagram, the first-order line terminates, for increasing negative g-values, in a critical point simultaneously with an increase of the separation between the two transitions. However, since the discontinuity in ψ decreases when the critical point is approached and since the discontinuity found in the Monte Carlo calculations for $g = -12$ is already close to the limit of our resolution, we have not found it fruitful to search for a more well-separated set of transitions by choosing a slightly larger negative value of g. For $g = -14.25$, we encounter only a single transition, and all quantities change smoothly in the transition region indicating that the transition is continuous.

The complete Monte Carlo information on the transition in the model is collected in the composite phase diagram shown in Fig. 3.2.14. In terms of the

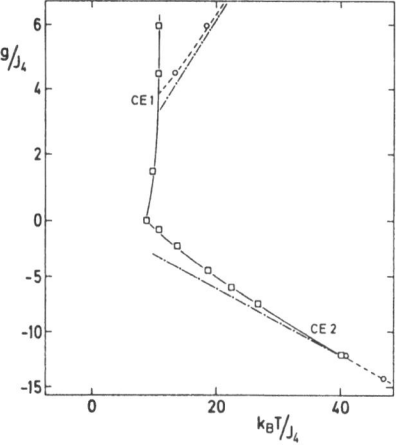

Fig. 3.2.14. Phase diagram (T, g) for the model in Eq. (3.2.14) as obtained from Monte Carlo calculations. The scaling field g is defined in Eq. (3.2.17). Squares denote first-order transitions and circles denote continuous transitions. The corresponding phase lines are given by solid and dashed lines. The approximate positions of the two critical end points $CE1$ and $CE2$ are also given. The dot-dash lines signify the pure pair interaction limit for the *fcc* lattice. (Cf. the mean-field phase diagram in Fig. 3.2.11.)

variables (T, g), the diagram has a kink for $g = 0$. When using the scaled variables in Eq. (3.2.17), this kink disappears and the phase boundary varies smoothly through $g = 0$. The various phase lines are shown and the approximate positions of the two critical end points, $CE1$ and $CE2$, are indicated. The position of $CE2$ is of course rather uncertain, $-7.5 \lesssim g(CE2) \lesssim -12$. The dot-dash lines in the figure correspond to the limit of $J_2^{\pm}/J_4 \to \infty$, i.e. pure pair interactions on an fcc lattice (Fisher and Burford 1967). This limit seems to have been reached effectively for $g \lesssim -14.25$. When comparing with the mean-field phase diagram in Fig. 3.2.11, it is seen that the mean-field theory in a quantitatively correct way describes the phase diagram for this model. Also, the overall structure of the diagram is in agreement with renormalization group calculations. Although not very precise, the Monte Carlo results for the position of the critical end points in Fig. 3.2.14 accord with the renormalization group prediction for the amplitude ratio, $g(CE2)/g(CE1) = -3 + O(\epsilon)$ (Kerszberg and Mukamel 1981).

3.3 Thermodynamics of One-dimensional Heisenberg Models

3.3.1   One-dimensional magnetic models

There exists a rigorous argument that no one-dimensional system with short-range forces of finite strength can sustain long-range ordering at finite temperatures (Stanley 1971). This argument is most easily illustrated by considering a completely ordered ferromagnetic Ising chain with N spins of quantum number S. The cost in internal energy of reversing a single spin is $4JS^2$. Since any one of the N spins can be reversed, the entropy associated with such a spin reversal is $k_B T \ln N$, and hence the change in free energy is $\Delta F = 4JS^2 - k_B T \ln N$. For a large system $(N > \exp(4JS^2/k_B T))$, ΔF is negative for all finite temperatures. Hence, the ordered state will break up into domains and the long-range order will disappear. For Heisenberg chains with a continuous symmetry, a similar argument can be advanced that the spin-wave dynamics abolishes long-range order.

Consequently, phase transitions and critical phenomena do not take place in one-dimensional systems and the thermodynamics is completely controlled by fluctuations.[*] This is not to say that one-dimensional systems are only dominated by disorderly processes. On the contrary, one-dimensional systems exhibit distinct structures in the dynamical behavior and sustain well-defined excitations. This kind of orderly behavior is mirrored in the thermodynamics.

To a large extent, it remains true that the theoretical physicists' preoccupation with one-dimensional models within statistical mechanics is governed by the higher mathematical tractability offered by low-dimensional problems. However, in recent years it has turned out that there exists a wide selection of real physical materials and phenomena which behave in effect as one-dimensional systems (see e.g. Bernasconi and Schneider 1981).

---

[*] This entails that mean-field theory is completely useless in one dimension.

A resurging interest in one-dimensional statistical physics has developed recently through the discovery that the dynamic as well as static properties of these systems may be significantly influenced by their capacity of supporting non-linear excitations (or solitons) (Krumhansl and Schrieffer 1975). Now, a vast and rapidly increasing amount of literature has emerged on this subject (for recent reviews, see e.g. Maki 1982).

We shall here be concerned with one-dimensional magnetic models. A few exact solutions are available for linear chains, notably the Ising model (Ising 1925), the spin$-\frac{1}{2}$ XY model (Lieb et al. 1961), and the classical Heisenberg model (Fisher 1964). Scalapino et al. (1972) have pointed out that all classical statistical mechanical models (e.g. including anisotropies and symmetry-breaking fields) can be solved exactly by numerical transfer matrix techniques. Here, we want to consider anisotropic ferromagnetic classical Heisenberg models in symmetry-breaking fields defined by the discrete lattice spin-Hamiltonian

$$H = -J \sum_i \bar{S}_i \cdot \bar{S}_{i+1} + A \sum_i (S_i^z)^2 - g\mu_B H \sum_i S_i^x. \tag{3.3.1}$$

$A > 0$ denotes the strength of a single-site crystal-field anisotropy, $g\mu_B$ is the magnetic moment of the classical spin vector, $\bar{S}_i = (S_i^x, S_i^y, S_i^z)$, and H is a magnetic field applied in the (x-y)-plane perpendicular to the anisotropy axis. The magnetic field breaks the rotational symmetry of the spin components in the (x-y)-plane. As we shall discuss in Sec. 3.3.4, there exists a number of physical realizations of the model in Eq. (3.3.1). Mikeska (1978) has pointed out that at low temperatures, $k_B T < (JA)^{\frac{1}{2}}$, where the spins are almost confined to the (x-y)-plane, the model in Eq. (3.3.1) may be mapped onto the continuous sine-Gordon model

$$H_{sG} = \int dx [\frac{1}{2} J (\frac{d\theta}{dx})^2 - g\mu_B H (\cos\theta - 1)], \tag{3.3.2}$$

where θ is the polar angle of the planar spins. Since the sine-Gordon model is a soliton-bearing system, it has been argued that Eq. (3.3.1) will also support solitons at low temperatures. Recent work however, has cast doubt on the validity of the mapping between the two models (Loveluck et al. 1980, Kumar and Samalam 1982).

The transfer matrix problem has been worked out for Eq. (3.3.1) in a number of limiting cases, e.g. $A = 0$ (Blume et al. 1975), $A = \infty$ (Patkós and Ruján 1979), and $H = 0$ (Loveluck et al. 1975). So far, the general case of $A \neq 0$ and $H \neq 0$ has not been considered by transfer matrix techniques. However, Schneider, Stoll, and coworkers have performed a number of numerical molecular dynamics calculations on the general model (see e.g. Loveluck et al. 1982 and references therein). Their activity has predominantly focused on dynamical aspects of the model, such as dynamic structure factors and time-dependent correlation functions.

Surprisingly little numerically exact information is available on the static thermodynamic properties of the general model in Eq. (3.3.1). This is particularly true of static fluctuation quantities, such as specific heat and susceptibilities. The need for such information as a means for comparison with experimental measurements has been emphasized within the last few years by the appearence of specific heat measurements for linear-chain magnets (Ramirez and Wolf 1982, Borsa et al. 1983). In Sec. 3.3.2, we shall present the results of a Monte Carlo simulation study of the general model in Eq. (3.3.1) with special attention paid to the specific heat as a function of T, A, and H.

A computer simulation approach to calculate the thermodynamics of linear magnetic models shares a drawback with the numerically exact transfer matrix approach and, to some extent, also with analytically exact solutions. This drawback is that no physical picture enters which models the characteristic excitation modes of the system. This is particularly severe in relation to questions about solitons which can only indirectly be investigated by these approaches.

3.3.2   The anisotropic Heisenberg model in a magnetic field

A computer simulation study of the classical anisotropic Heisenberg model in a field has been carried out by Mouritsen et al. (1984). We shall report on this work in the present section. Focus will be on the ferromagnetic case, $J > 0$, and a few remarks on the antiferromagnetic case will be offered towards the end of Sec. 3.3.4.

The Monte Carlo calculations are carried out on cyclic chains with N classical spins. The chains of interacting spins are brought to thermodynamic equilibrium by means of sequential visitation of lattice sites in combination with a Glauber single-site excitation mechanism. Each excitation rotates the spin through a random solid angle. The specific heat, $C_H(T)$, is derived via the fluctuation theorem, Eq. (2.1.7). The main part of the calculations is performed on chains with $N = 100$ spins. Calculations on longer chains with $N = 400$ spins show that finite-size effects can be neglected for the $N = 100$ spin chains as far as C_H is concerned.

Despite the lack of true critical fluctuations in one-dimensional spin systems, it turns out to be extremely demanding to obtain accurate numerical values of C_H in the presence of the field. This is due to a very delicate competition between the ordering effect of the magnetic field term and the tendency of the entropic part of the free energy to destroy long-range ordering. Accordingly, the specific heat results reported below are based on very extensive statistics corresponding to ensembles of about $40000N$ microconfigurations. Moreover, the final values of C_H are obtained by averaging over five to ten different ensembles constructed by using different Markov chains. Thus, the statistics required for the present calculations is about two orders of magnitude larger than that needed to calculate C_H in the neighborhood of an ordinary three-dimensional critical point (cf. Sec. 3.1.2 and Fig. 3.1.3). We believe that similar unusual demands are responsible for the difficulties encountered by Gerling and Landau (1982) in their attempt to calculate C_H for the classical XY chain in a magnetic field. These authors generated 2500 Monte Carlo steps per site for chains with $N = 2000$ sites. For the general model in Eq. (3.3.1), it turns out that statistics involving $2 - 4 \times 10^5 N$ microconfigurations yields C_H with an accuracy of a few per cent. This amount of statistics will

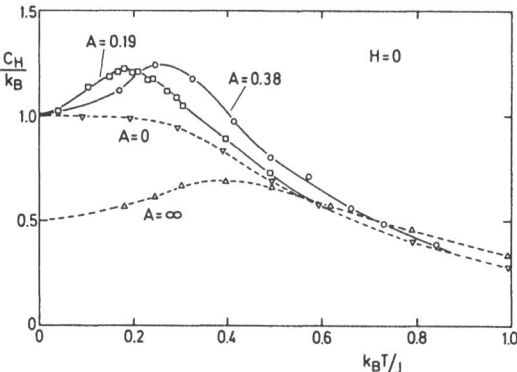

Fig.3.3.1. Specific heat, C_H/k_B, as a function of temperature for the one-dimensional anisotropic Heisenberg ferromagnet in zero field, Eq. (3.3.1). The anisotropy A is given in units of J. Monte Carlo data are denoted by open symbols and solid lines to guide the eye. The exact results for the isotropic Heisenberg model and the planar XY model are given by dashed lines.

usually determine the internal energy and the field-induced magnetization with an accuracy of about 0.05% and 0.5% ,respectively. For very low values of the field, the magnetization is less well determined. It is a general observation that the most extensive statistics is needed at very low temperatures (where the sampling scheme chosen is less efficient) and for those values of temperature and field where C_H displays a peak. Since in the following we shall be interested in locating such peaks which are rather broad, specific heat data of unusually high quality are required.

In order to check the reliability of the computer simulations, we have made a comparison with the exact analytical solution for the zero-field isotropic chain ($H = A = 0$) (Fisher 1964) and with the numerically exact transfer matrix results for the planar XY chain in a field ($A/J \rightarrow \infty$) (Patkós and Ruján 1979). In both cases, we obtain excellent agreement.

Figure 3.3.1 gives a selection of Monte Carlo data and exact results of the zero-field specific heat for various values of the anisotropy. First of all, this figure shows that the Monte Carlo results are identical with the exact results in the two limiting cases, $A = 0$ and $A = \infty$. The specific heat for the isotropic chain decreases monotonously as the temperature is raised. When a finite anisotropy is introduced, C_H develops a rounded peak which becomes broader and moves towards higher temperatures as the anisotropy is increased. The models with $0 \leq A < \infty$ have indentical low- and high-temperature limit behavior which is dictated by the dimensionality of the spin vectors. In contrast, the planar model, which has the spins confined to the (x-y)-plane, has a different behavior in these limits. In particular, $C_H(T = 0) = \frac{1}{2}k_B$ since each planar spin has only one degree of freedom. As a function of anisotropy, the peak intensity of $C_H(T)$ on the temperature axis also displays a maximum.

The effect of turning on a uniform magnetic field parallel to the easy plane is a shifting of the peaks in Fig. (3.3.1) towards higher temperatures. For fixed anisotropy and as a function of the magnetic field, the peak intensity of $C_H(T)$ on the temperature axis also displays a maximum.

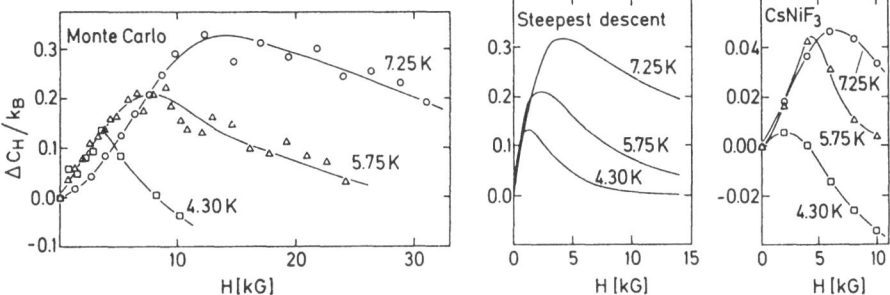

Fig. 3.3.2. Specific heat difference, $\Delta C_H = C_H - C_{H=0}$, as a function of magnetic field for three different temperatures. The field and temperature are given in absolute units pertinent to $CsNiF_3$ (cf. Eq. (3.3.7)). The experimental results of Ramirez and Wolf (1982) for $CsNiF_3$ are compared with the results of theoretical calculations on the discrete lattice model in Eq. (3.3.1) (Monte Carlo) and the continuum model in Eq. (3.3.3) (Steepest descent, Eqs. (3.3.4) - (3.3.6)). The scatter in the Monte Carlo data is representative of the numerical accuracy. The model calculations correspond to an anisotropy of $A/J = 0.19$ ($\sim$ 4.5K for $CsNiF_3$).

Rather than attempting to describe quantitatively the specific heat peak as a function of A, T, and H, we shall concentrate on a few cross-sections of this parameter space characterized by fixed values of A and T. It is of particular interest to compare the information on the field ·dependence of C_H for specific values of A and T with experimental measurements on linear-chain magnets (Sec. 3.3.4).

In Fig. 3.3.2 is shown selected Monte Carlo results for the field dependence of C_H at different temperatures for $A/J = 0.19$. In this figure, the contribution to C_H due to the magnetic field is isolated by plotting the specific heat difference, $\Delta C_H = C_H - C_{H=0}$. In order to facilitate the comparison with experimental

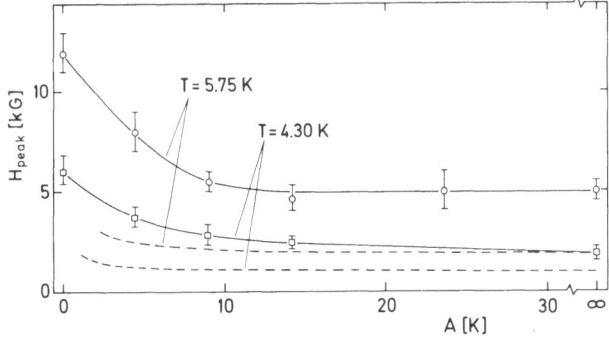

Fig. 3.3.3. Position, H_{peak}, on the field axis of the maximum of the specific heat plotted *vs* anisotropy for two different temperatures. Field, temperature, and anisotropy are given in absolute units pertinent to $CsNiF_3$ (cf. Eq. (3.3.7)). Results are given from theoretical calculations on the discrete lattice model in Eq. (3.3.1) (Monte Carlo, open symbols and solid guide lines) and the continuum model in Eq. (3.3.3) (Steepest descent, Eqs. (3.3.4) - (3.3.6), dashed lines). The theoretical curves do not extend down to zero anisotropy due to the fact that the singularity at $2A/g\mu_B H = 3$ of Eq. (3.3.6) intervenes before the specific heat has attained its maximum.

measurements on a specific substance, $CsNiF_3$ (Sec. 3.3.4), we give the field and the temperature in absolute units, cf. Eq. (3.3.7). From Fig. 3.3.2, we obtain the striking result that C_H exhibits a pronounced broadened peak as the field is varied. For increasing temperatures, the peak broadens, increases in intensity, and moves towards higher fields. A similar overall behavior is found for other values of the anisotropy.

The anisotropy dependence of the position, H_{peak}, of the specific heat maxima on the field axis is shown in Fig. 3.3.3 for two different temperatures. The various quantities are expressed in absolute units pertinent to $CsNiF_3$. For all temperatures, it is found that the specific heat maximum moves towards lower fields as the anisotropy is increased. From Fig. 3.3.3, it appears that the model for the two temperatures considered in effect behaves as a planar XY model for $A/J \gtrsim 0.6$.

3.3.3 Comparison with theoretical calculations on a continuum model

The major part of approximate analytical calculations on one-dimensional magnetic models are performed on continuum models where a continuous spin field, $\bar{S}(\bar{r})$, takes the role of the discrete lattice spins, $\bar{S}(\bar{r}_i)$. For most models, the continuum version represents a problem of higher mathematical tractability. However, many questions then arise as to how well a continuum model may describe a discrete spin system. Obviously, at very low temperatures where long-wavelength fluctuations prevail, continuum and discrete spin models should lead to the same results. But what are »low temperatures« for specific values of the model parameters? A second very delicate question is related to how finite-lattice-spacing effects behave when the continuum limit is taken.

Here, we shall use the Monte Carlo simulation results presented in Sec. 3.3.2 for the discrete model in Eq. (3.3.1) to attempt an answer to the first question posed above through a comparison with theoretical calculations on the continuum version of Eq. (3.3.1):

$$H = \int dx \{ \tfrac{1}{2} J [\frac{d\bar{S}}{dx}]^2 + A(S^z)^2 - g\mu_B H(S^x - 1) \}. \tag{3.3.3}$$

Obviously, Eq. (3.3.3) reduces to the sine-Gordon model, Eq. (3.3.2), in the limit of extreme anisotropy, $(A/J)^{\frac{1}{2}} \gg k_B T$.

The low-temperature thermodynamics of the model in Eq. (3.3.3) has recently been calculated via a steepest-descent approach by Fogedby et al. (1983) and by Leung and Bishop (1983). These low-temperature analytical calculations are based on the physical picture that the spin system at low fields forms a stable gas of spin waves and domain walls (solitons). The result for the field-induced specific heat contribution, ΔC_H, is given by

$$\Delta C_H / k_B = (\frac{E_{DW}}{k_B T})^2 n_s, \tag{3.3.4}$$

where $E_{DW} = 8(Jg\mu_B H)^{\frac{1}{2}}$ is the domain-wall energy and n_s is the soliton density

$$n_s = (8\pi)^{-\frac{1}{2}} (\frac{E_{DW}}{J})(\frac{E_{DW}}{k_B T})^{\frac{1}{2}} \bar{A}(\frac{2A}{g\mu_B H}) \exp(-\frac{E_{DW}}{k_B T}). \qquad (3.3.5)$$

The anisotropy function is defined as

$$\bar{A}(x) = \frac{(1 + [1 + x]^{\frac{1}{2}})(2 + [1 + x]^{\frac{1}{2}})}{(x - 3)^{\frac{1}{2}} x^{\frac{1}{2}}} \qquad (3.3.6)$$

which reduces to unity in the sine-Gordon limit. Hence $\bar{A}$ accounts for the out-of-plane fluctuations.

The results derived from the theoretical expression, Eq. (3.3.4), are also shown in Fig. 3.3.2. We note a qualitative accordance with the Monte Carlo results for the discrete model. Still, the peak positions are markedly different. The amplitudes, however, are very similar. The theoretical predictions for the anisotropy dependence of H_{peak} for various temperatures are compared with the Monte Carlo results in Fig. 3.3.3. This figure shows that the marked difference between H_{peak} for the two sets of calculations persists for all values of A. However, in accordance with expectations, the relative deviation decreases when the temperature is lowered.

The question naturally arises to which extent solitons cause the specific heat peaks in Fig. 3.3.2. Lacking an operative and unambiguous definition of a solitary spin wave in a discrete spin chain, we are unable to deliver a definite answer to this question by the present type of Monte Carlo calculations. The analytical theories leading to Eqs. (3.3.4) – (3.3.6) relate the specific heat maximum to the structure of the spin-wave-renormalized or »dressed« domain-wall density in Eq. (3.3.5). The validity of these theories is restricted to low temperatures, $k_B T \ll (Jg\mu_B H)^{\frac{1}{2}}$, and the question remains open whether the domain walls of the model in Eq. (3.3.3) are distinct or masked by spin-wave excitations for the range of temperatures considered here. That the specific heat maxima are not necessarily unique features of soliton-bearing systems is perhaps most clearly indicated by the fact that we also find a broadened maximum in the Monte Carlo C_H for the isotropic ($A = 0$) chain in a field. This model is spin-wave dominated at low temperatures and does not support static finite-energy domain walls (see e.g. Tognetti et al. 1983).

The temperature dependence of the position, H_{peak}, of the specific heat maxima in Fig. 3.2.2 is analyzed in Fig. 3.3.4 in which H_{peak} is plotted as a function of T^2. Results derived from Eqs. (3.3.4) – (3.3.6) as well as from the sine-Gordon model are also included in Fig. 3.3.4. The sine-Gordon model leads to a linear relationship between H_{peak} and T^2 (Currie et al. 1980). In contrast, the analytical calculation on the finite-anisotropy continuum model in Eq. (3.3.3) gives a non-linear dependence of H_{peak} on T^2 due to the presence of the anisotropy function in Eq. (3.3.6). The Monte Carlo results in Fig. 3.3.4 also suggest a non-linear behavior. Still, there is a large quantitative discrepancy between the numerically exact computer simulation results for the discrete model and the low-temperature analytical calculation on the continuum model. In accordance with expectations, the relative difference between the two sets of model calculations decreases as the temperature is lowered.

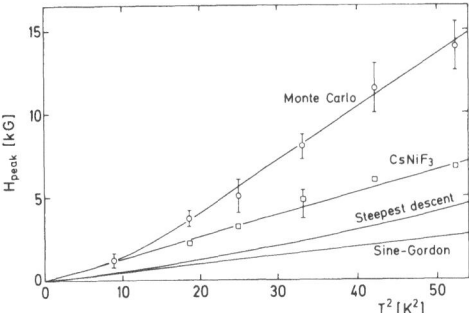

Fig. 3.3.4. Position, H_{peak}, on the field axis of the maximum of the specific heat plotted *vs* temperature squared. The field and temperature are given in absolute units pertinent to $CsNiF_3$ (Eq. (3.3.7)). The experimental results of Ramirez and Wolf (1982) are compared with the results of three different model calculations. The curves labeled steepest descent and Monte Carlo are derived from calculations on the continuum model in Eq. (3.3.3) and the discrete lattice model in Eq. (3.3.1), respectively. Both of these model calculations correspond to an anisotropy of $A/J = 0.19$ ($\sim$ 4.5K for $CsNiF_3$).

To conclude the comparison between the continuum and discrete lattice model versions of the anisotropic Heisenberg chain in a field it may be stated that, even down to the lowest temperatures investigated in the Monte Carlo simulations, $k_BT/J \approx 0.18$ (4.3K for $CsNiF_3$), the continuum model does not provide a very accurate quantitative description of the thermodynamics. The marked discrepancy calls for an investigation of finite-temperature and finite-lattice-spacing effects.

Finally, a remark is in order on the anomaly predicted by the steepest-descent approach (Fogedby et al. 1983, Leung and Bishop 1983) to be present in the thermodynamic properties at a »critical« field $g\mu_B H = 2A/3$ ($\sim$ 18kG for $CsNiF_3$). This anomaly, caused by the singular structure of the anisotropy function, Eq. (3.3.6), at $x = 3$, is a spurious feature of the steepest-descent calculation which only takes into account static domain walls. At the »critical field«, the static and dynamic domain walls will couple leading to a smearing of the singularity (Magyari and Thomas 1983, Mikeska and Osano 1983). This is supported by the Monte Carlo calculations which show no sign of anomalies around the »critical field«, neither for the specific heat nor for any other bulk thermodynamic quantity.

3.3.4 A model of the linear magnet $CsNiF_3$?

The magnetic salt $CsNiF_3$ has a crystal structure which may be described as chains of $Ni^{++}F_3^-$ ions magnetically isolated from each other by the Cs^+ ions (Mikeska 1979). The quantum spin-1 Ni^{++} ions interact within the chain via ferromagnetic exchange interactions. The ratio between the inter- and intrachain interactions is of the order of 10^{-3}. Hence, for temperatures above the Néel point, $T_N = 2.61$K, the chains are effectively decoupled from each other and the system is in effect a linear magnet. The individual chain is furthermore subject to a single-ion anisotropy which confines, at sufficiently low temperatures, the spins within the easy plane perpendicular to the chain axis. This led Mikeska (1978) to suggest $CsNiF_3$ in an in-plane field as a possible candidate for a physical realization of a system which is described by the sine-Gordon model and which may support non-linear excitations (solitons). Now, it is generally believed that the magnetic excitations of $CsNiF_3$ are rather described by the discrete classical spin Hamiltonian in Eq. (3.3.1) with the following set of parameters[*]

$$J/k_B = 23.6K, \quad A/k_B = 4.5K, \quad g\mu_B/k_B = 0.16K kG^{-1}. \qquad (3.3.7)$$

(Kjems and Steiner 1978; for recent reviews, see Mikeska 1979 and Steiner et al. 1983).

Recently, Ramirez and Wolf (1982) have measured the specific heat of $CsNiF_3$ in an applied field. The experimental results are reproduced in Fig. 3.3.2. In the absence of specific heat calculations for the model in Eq. (3.3.1), Ramirez and Wolf compare their experimental results with the theoretical predictions for the classical sine-Gordon model (Currie et al. 1980). They find qualitative agreement and therefore attribute the peak found in the specific heat as a function of the field to the characteristic non-linear domain-wall excitations of the sine-Gordon model. Renormalizing the domain-wall energy in a somewhat *ad hoc* manner, they even obtain a quantitative fit.

Having available the Monte Carlo results for the finite-anisotropy model in Eq. (3.3.1), we are now in a much better position to evaluate the usefulness of the classical spin model as a model of the field-dependent thermodynamics of $CsNiF_3$. Figure 3.3.2 clearly shows that the classical model in Eq. (3.3.1) does not pass the test as a proper model for the specific heat of $CsNiF_3$. Although there is qualitative agreement between experiment and model calculation, there is a severe quantitative discrepancy with respect to the peak positions as well as to the intensities, the latter being almost an order of magnitude smaller for $CsNiF_3$. The scaling of the experimental peak position, H_{peak}, with temperature squared is investigated in Fig. 3.3.4. As suggested by the Monte Carlo simulations and by the steepest descent calculations, this scaling is not linear. Since the experimental data cannot be distinguished from a linear function in the temperature interval ranging from 4K to 7K, the non-linear prediction can only be tested by measurements in a wider temperature range.

These findings strongly suggest that quantum effects in $CsNiF_3$ may be more important than hitherto expected in the temperature range considered in the experiments. This suggestion is supported by recent numerical quantum calculations on the planar model in a field (Chui and Ma 1983) and on the model given in Eq. (3.3.1) (Schneider and Stoll 1982). Thus, we conclude that quantum effects are essential to include in a proper model for the thermodynamics of $CsNiF_3$.

There exists another well-known linear-chain magnet, $TMMC [(CH_3)_4 NMnCl_3]$, in which quantum effects may be less dominant because of the higher spin quantum number of the magnetic Mn^{++} ions, $S = \frac{5}{2}$ (Mikeska 1979, Borsa et al. 1983). This linear magnetic system is believed to be well described by the antifer-

---

[*] Quantum corrections are expected to increase A to $A/k_B = 9K$.

romagnetic version of Eq. (3.3.1) with the single-site anisotropy replaced by a pair interaction anisotropy, $-\delta S_i^z S_{i+1}^z$, with $\delta = 0.016$. Results of preliminary Monte Carlo calculations[*] on the classical version of this model have been compared with the specific heat measurements by Borsa et al. (1983) as a function of temperature and magnetic field parallel and perpendicular to the chain axis. The result is that the quantum effects as mirrored in C_H are much less dominant for $TMMC$ than for $CsNiF_3$.

---

[*] O.G. Mouritsen, H. Jensen, and H.C. Fogedby, unpublished results.

4. Testing Modern Theories of Critical Phenomena

The modern theories of critical phenomena are based on a variety of hypotheses, assumptions, and specific approximations which may enter a particular theory at many different stages of its development.

By nature, a *hypothesis* is often general and concerned with a complex of problems. Its usefulness is determined by its success in aiding a theory to generate reliable and accurate predictions. As an example, we may mention the hypothesis, underlying the whole renormalization group theory, which states that critical phenomena are associated with fixed points of the renormalization group transformations. Obviously, such a hypothesis cannot be given an overall verification. However, specific implications of the hypothesis can be tested, and the outcome of such tests may give testimony to the usefulness of the hypothesis. In this chapter, we shall be concerned with a number of specific predictions of the renormalization group approach to critical phenomena. The validity of these predictions will be tested by computer studies of appropriate microscopic models. In Sec. 4.1, we examine the renormalization group prediction that lack of stable fixed points implies fluctuation-induced first-order phase transitions. Furthermore, tests will be performed of the theoretical prediction that finite symmetry-breaking fields are required to restore continuous transitions. In Sec. 4.2, a more intimate examination of the renormalization group approach is performed by a numerical computer study at marginal spatial dimensionality. The concept of a marginal dimension lies at the roots of the whole theory. The prediction is that the critical behavior is essentially that given by mean-field theory. Since this prediction does not rest on any approximate ϵ- and $\frac{1}{n}$-expansion, a study of critical behavior at marginal dimensionality is a study at the core of the renormalization group approach.

Assumptions are often introduced into theories as informed guesses of unknown functional relationships. The assumptions may be well-founded conjectures or they may simply be introduced *ad hoc* in order to bring the theory on a tractable form. Usually, *ad hoc* assumptions are justified *a posteriori*. Assumptions can in principle be verified or falsified, e.g. by computer simulation. If they turn out to be wrong, they may in many cases still be useful as *approximations*. In Sec. 4.3, we focus on such a situation by studying an analytical theory of the phase transition in three-dimensional Ising models. This theory is based on an *ad hoc* assumption about the behavior of certain correlation functions at the critical point.

4.1 Fluctuation-induced First-order Phase Transitions

4.1.1   The role of fixed points in the renormalization group theory

The starting point in a Wilson-type renormalization group calculation of the critical behavior of a physical system is the Landau-Ginzburg-Wilson Hamiltonian, H_{LGW} (Fisher 1974). This Hamiltonian, which is basically a generalized Landau expansion in terms of the various symmetry-breaking order parameters and in terms of a number of coupling constants, is constructed on the basis of symmetry considerations exclusively (Mukamel and Krinsky 1976a). By applying iteratively an appropriate renormalization group transformation to H_{LGW} in $d = 4 - \epsilon$ dimensions, one studies the flow pattern in the coupling parameter space. The fundamental hypothesis of the entire approach is then to associate stable fixed points in the flow diagram with critical behavior of the system under consideration. From the transformation properties in the neigborhood of the fixed points, universal critical parameters are derived.

In this chapter, the following two possibilities will be considered: (i) there exist no stable fixed points of the renormalization group equations, and (ii) stable fixed points exist but none of these are physically accessible in a flow from the initial H_{LGW}. For both these possibilities, the renormalization group flow will take the system into a region of the coupling parameter space where it becomes thermodynamically unstable (i.e. the fourth-order terms are no longer positive definite).

It has been suggested by Bak et al. (1976a) (see also Brazovsky and Dzyaloshinsky 1975, Brazovsky et al. 1976, and Alessandrini et al. 1976) that the lack of stable fixed points implies that the system undergoes a first-order phase transition. The first-order phase transition takes place when the correlation length exceeds a certain limit at which fluctuations in the order parameter make a discontinuous transition energetically favorable. These so-called *fluctuation-induced first-order phase transitions* occur even in cases where the mean-field theory predicts continuous transitions. For systems which have $n \leq 3$ order parameter components, there is always one stable fixed point, the isotropic fixed point (Brézin et al. 1974). This leads immediately to the definition of the Ising, XY, and Heisenberg universality classes for $n = 1, 2$, and 3. However, certain phase transitions which involve a breaking of the translational symmetry are described by n-component vector models with $n \geq 4$ (Mukamel 1976, Alben 1974). This has lead to the discovery of several new universality classes (Mukamel and Krinsky 1976a,b, Bak and Mukamel 1976). In particular, it has been shown that the H_{LGW} appropriate for many spin models with $n \geq 4$ may possess no stable fixed points (Bak et al. 1976a, Löser and Sólyom 1978). Some physical realizations of such spin models are UO_2 $(n = 6)$, MnO $(n = 8)$, Eu $(n = 12)$, Cr $(n = 12)$, TbP $(n = 4)$ (Bak et al. 1976a), solid ^3He $(n = 18)$ (Bak and Rasmussen 1981), and CrN $(n = 12)$ (Mrozińska et al. 1979). Experiments on these materials confirme that the magnetic transitions are indeed of first order thus lending substantial support to the original suggestion that the lack of stable fixed points (within the ϵ-expansion) implies that continuous transitions are not allowed. For completeness, it should be pointed out that the existence of a stable fixed point of the renormalization group flow does not necessarily in itself guarantee that a continuous transition will take place. In fact, the initial H_{LGW} may very well be outside the region of attraction of the stable fixed points. MnS_2 is an example of a system with $n \geq 4$, which has a stable fixed point but undergoes a first-order transition (Hastings and Corliss 1976).

These last remarks lead us to consider in more detail the second possibility (ii) outlined above: the stable fixed points are not accessible. To be specific, we shall focus here on the following situation (other situations are discussed by Domany et al. 1977, Rudnick 1978, Kerszberg and Mukamel 1979, 1981, Blankschtein and Mukamel 1982): Consider a model with $n \geq 4$ which possesses no stable fixed points in $4 - \epsilon$ dimensions and therefore is expected to undergo a first-order phase transition. Let the model be subject to a symmetry-breaking field, such as a uniaxial stress or a magnetic field, which splits the degeneracy of the n components of the order parameter. If by this procedure the number of order parameter components of the ground state manifold is reduced below four, the isotropic fixed point becomes stable and a continuous transition is allowed. This situation was first studied by Bak et al. (1976b), who, as an example, pointed to the observed destruction of the first-order phase transition in MnO (Bloch et al. 1975, 1980) in the presence of a uniaxial stress along the [111]-direction (reducing n from 8 to 2).

The question naturally arises as to whether the crossover from first-order to continuous transitions takes place for a finite value of the symmetry-breaking field or whether an arbitrarily small field can destroy the first-order transition. By studying an $n = 6$ model appropriate for UO_2, Sólyom and Grest (1977) have argued that the crossover should occur at a finite field. This conclusion is supported by renormalization group calculations on models with $n \geq 4$ (Kerszberg and Mukamel 1981). A similar question can be posed for systems with $n \leq 3$ and no accessible fixed points. An example is $RbCaF_3$ subjected to a [100] uniaxial stress (lowering n from 3 to 2) where the theoretical analysis (Aharony and Bruce 1979) in accordance with experiments (Buzaré et al. 1979) predicts a crossover to continuous transitions for a finite stress.

In Secs. 4.1.2 - 4.1.5, computer studies of a variety of three-dimensional models undergoing fluctuation-induced first-order phase transitions will be described. Two-dimensional models will be discussed in Sec. 5.3.

4.1.2   Motivation for computer studies of fluctuation-induced first-order phase transitions

The entire theory of fluctuation-induced first-order phase transitions and the crossover to continuous transitions in symmetry-breaking fields is based on *pheno-menological* renormalization group arguments in $4 - \epsilon$ dimensions. In principle, such calculations could be misleading when extrapolated to three dimensions. For real magnetic materials, like UO_2 or MnO, the details of the microscopic interactions are very seldom known. It is therefore virtually impossible to prove rigorously that an appropriate mean-field theory, taking into account the actual physical interactions, might not eventually lead to a first-order transition irrespective of the fluctuations. If this were the case, the experimental findings of first-order transitions could not be taken as supporting the phenomenological renormalization group theory.

It is therefore essential to perform investigations on *precisely defined three-dimensional microscopic interaction models* in testing the ideas of the *phenomenological* renormalization theory in $4 - \epsilon$ *dimensions*. For precisely defined microscopic models, it is a trivial matter to check whether or not mean-field (or Landau) theory leads to a first-order transition.

Computer simulations as well as series analyses lend themselves to the calculation of phase diagrams for microscopic models. Usually, the location of a phase boundary is easy to determine using either technique. As for the nature of a specific phase transition, neither of the techniques can be used to prove the existence of a continuous transition. However, in many cases it is possible to show that the analysis is consistent with a transition being continuous. Computer simulation techniques can often provide very strong evidence for a first-order transition. It is much more difficult from analysis of series expansions to establish that a transition is of first order. The most common approach is to analyze the series assuming critical singularities. If the convergence is very irregular, or if high- and low-temperature series analyses lead to significantly different estimates of the transition temperature, this is often taken as supportive evidence of a first-order transition. Nevertheless, it can never be excluded that the asymptotic series analysis may suffer from finite-series effects or that unexpected non-physical singularities interfere. Furthermore, for many models of physical interest only high-temperature series are available, and for the majority of models only very short series have been derived.

In contrast, both the high- and low-temperature regimes are accessible by computer simulation methods. Moreover, very complicated models may be treated without introducing any new techniques. This is a considerable advantage when studying fluctuation-induced first-order phase transitions in antiferromagnets with many-component order parameters, systems which usually require microscopic models of considerable complexity. Obviously, computer simulation is an ideal technique for exploring fluctuation-induced first-order phase transitions.

Renormalization group calculations, series analyses, and computer simulations each have their advantages and drawbacks. When these methods in combination are applied to a particular problem, they complement one another and constitute today's most powerful theoretical approach to the field of phase transitions. A verification of this statement will be attempted in the following sections.

4.1.3   Phase transitions in antiferromagnets with order parameters of dimension $n = 6$ and $n = 3$

The earliest computer simulation study of a fluctuation-induced first-order phase transition is that performed by Mouritsen et al. (1977) who studied an $n = 6$ model pertinent to UO_2 and $NdSn_3$. To provide a comparison with a continuous transition, these authors also studied a closely related model with $n = 3$ (Mouritsen et al. 1978) pertinent to $RbMnF_3$.

The two models are defined on simple cubic lattices with classical spins and have the same form of the Hamiltonian as given by

$$H = \sum_{\substack{j,k\,(\neq j)}}^{nn} J_1(\bar{r}_{jk})\,\bar{S}_j \cdot \bar{S}_k + \sum_{\substack{j,k\,(\neq j)}}^{nn} K(\bar{r}_{jk})\,(\bar{r}_{jk} \cdot \bar{S}_j)(\bar{r}_{jk} \cdot \bar{S}_k)/r_o^2$$

$$+ J_2 \sum_{\substack{j,k\,(\neq j)}}^{nnn} \bar{S}_j \cdot \bar{S}_k + P\sum_j (S_{jx}^4 + S_{jy}^4 + S_{jz}^4). \tag{4.1.1}$$

nn and nnn indicate that the summations are extended over the six nearest neighbor pairs and twelve next-nearest neighbor pairs, respectively. $\bar{r}_{jk}$ is the vector connecting the jth and kth spin, and r_o is the lattice parameter of the cubic unit cell. $J_1(\bar{r}_{jk})$, J_2, $K(\bar{r}_{jk})$, and P are model parameters. We shall in this section assume the coupling parameters to be directionally independent, $J_1(\bar{r}_{jk}) = J_1$ and $K(\bar{r}_{jk}) = K$, and choose the parameters in the following way

$$P = K/5 = -2J_2 = J_1 \equiv J < 0 \tag{4.1.2}$$

for the $n = 6$ model. The $n = 3$ model only differs from the $n = 6$ model by the replacement of K by $-K$. Since $P < 0$, the single-ion term tends to direct the spins along the principle axes of the unit cell. This facilitates the measurement of the order parameter in the Monte Carlo experiments.

At low temperatures, the choice of model parameters leads to type-III simple cubic antiferromagnetic structures as shown in Fig. 4.1.1. For both models, the structures are characterized by *three* propagation vectors

$$\bar{k}_1 = \frac{2\pi}{r_o}(\tfrac{1}{2},0,0), \quad \bar{k}_2 = \frac{2\pi}{r_o}(0,\tfrac{1}{2},0), \quad \bar{k}_3 = \frac{2\pi}{r_o}(0,0,\tfrac{1}{2}). \tag{4.1.3}$$

For the $n = 6$ model, the stable structures are of a transversal type with the sublattice magnetization in *two* equivalent directions perpendicular to the propagation

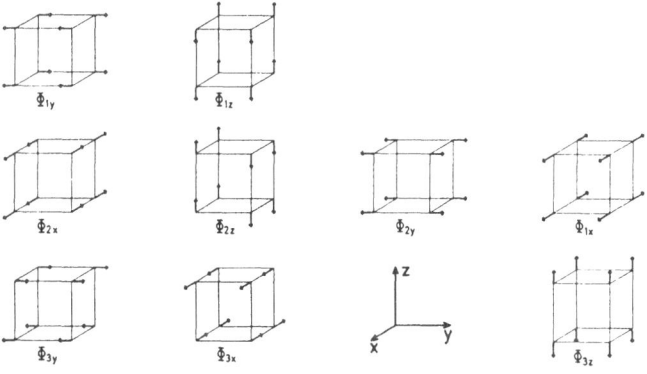

Fig. 4.1.1. Type-III simple cubic antiferromagnetic structures. The order parameter components, $\Phi_{i\alpha}$, are labeled by $i = 1, 2, 3$, corresponding to the propagation vectors of Eq. (4.1.3), and by $\alpha = x, y, z$, denoting the component of the spin vectors.

vectors. Thus the order parameter has the six components, Φ_{1y}, Φ_{1z}, Φ_{2x}, Φ_{2z}, Φ_{3x}, and Φ_{3y}, as shown in Fig. 4.1.1. The order parameter components are labeled $\Phi_{i\alpha}$, where $i = 1, 2, 3$ refers to the propagation vector, and $\alpha = x, y, z$ denotes the relevant component of the spin vectors. For the $n = 3$ model, the stable structures are of a longitudinal type with the sublattice magnetization along the propagation vectors. This leads to only three order parameter components, Φ_{1x}, Φ_{2y}, and Φ_{3z}.

The $n = 6$ model presented above was constructed within the spirit of the universality hypothesis. No attempt was made to correlate the form of the interactions included in the model to realistic interactions in a real physical system. Only the variables relevant for the universal classification of phase transitions are chosen so as to model magnetic materials such as UO_2 and $NdSn_3$. The magnetic structure of the $n = 6$ model is that observed for $NdSn_3$ (Lethuillet et al. 1973) and it is similar to the magnetic structure of UO_2 (Frazer et al. 1965) except that the spins in UO_2 are situated in a face-centered cubic lattice. However, it turns out that the Landau-Ginzburg-Wilson Hamiltonians are the same for the UO_2 structure, for the $NdSn_3$ structure, and for the $n = 6$ model. Consequently, they all belong to the same universality class and the same critical behavior is expected. It is important to emphasize that to be sure that a given microscopic model belongs to a particular universality class it is *not* a sufficient condition that the microscopic model leads to the correct order parameters. It is also required that the microscopic model does not possess any accidental symmetries which cause some of the necessary fourth-order terms in the associated Landau-Ginzburg-Wilson Hamiltonian to vanish. This is a very delicate point which has been discussed by Knak Jensen et al. (1979) in the context of the $n = 6$ model.

The Landau-Ginzburg-Wilson Hamiltonian of the $n = 6$ model has no stable fixed points in $4 - \epsilon$ dimensions. On the contrary, a stable fixed point is accessible for the $n = 3$ model. The mean-field theory is *identical* for the two models and predicts continuous transitions. Thus we have a rather favorable set-up for examining whether or not the lack of stable fixed points in $4 - \epsilon$ dimensions implies a fluctuation-induced first-order transition in a three-dimensional model.

The Monte Carlo simulations on the $n = 6$ and $n = 3$ models focus on calculating the internal energy and the various order parameter components

$$\Phi_{i\alpha}(T) = N^{-1} < \sum_{j=1}^{N} S_{j\alpha} e^{i\bar{k}_i \cdot \bar{r}_j} >; \quad i = 1, 2, 3; \quad \alpha = x, y, z. \tag{4.1.4}$$

The main part of the calculations is carried out on a lattice with $N = 14^3$ sites. Lattices with $N = 8^3$ and 10^3 are also studied in order to investigate finite-size effects. The large number of order parameter components makes distribution functions for each component, $\Phi_{i\alpha}$, extremely useful in the transition region. For the $n = 6$ model in the transition region, small changes in temperature often cause the system to evolve into transient states characterized by long-range order in more than one of the order parameter components. The transient states persist for several thousand Monte Carlo steps per site but eventually they are damped and disappear. We interpret this behavior to be a consequence of the slow domain-growth kinetics of systems with $n > d$ (Safran 1981; cf. Sec. 5.4). When the system softens in the transition region, the order parameter fluctuations, following a change in temperature, may induce transitions from a homogeneous ordered state into inhomogeneous non-equilibrium states. These states are characterized by a mosaic structure composed by domains of different types of ordering. The relaxation out of such states is expected to be slow for topological reasons. No such behavior is observed for the $n = 3$ model. The occurrence of transient states for the $n = 6$ model necessitates extensive statistics, and for several temperatures in the transition region more than 10^4 Monte Carlo steps per site are required to obtain reliable ensemble averages.

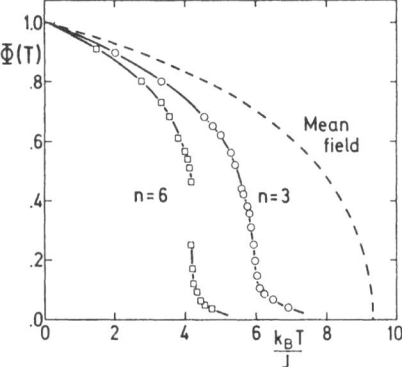

Fig. 4.1.2. Order parameter, $\Phi(T)$, vs temperature as obtained from Monte Carlo calculations on the $n = 6$ and $n = 3$ models arrayed on lattices with $N = 14^3$ sites. The mean-field order parameter curves are identical for the two models.

The results for the temperature dependence of the order parameter for the two models are shown in Fig. 4.1.2 together with the common mean-field prediction. We notice pronounced quantitative as well as qualitative differences between the results for the two models. The major difference is a discontinuous change in $\Phi(T)$ for the $n = 6$ model in contrast to a continuous variation of $\Phi(T)$ through the transition region for the $n = 3$ model. The discontinuity of $\Phi(T)$ is wiped out by fluctuations in the smaller lattices. These results immediately suggest that the $n = 6$ model undergoes a first-order transition whereas the $n = 3$ model exhibits a behavior consistent with a continuous transition. Obviously, the mean-field theory is quantitatively incorrect for both models and is moreover qualitatively wrong for the $n = 6$ model. Further evidence of a first-order transition in the $n = 6$ model is provided by Fig. 4.1.3 which shows the variation of the internal energy and the order parameter in the transition region. A clear hysteresis behavior is observed from increasing and decreasing temperature scans. The hysteresis extends over a very narrow temperature region, $\Delta T / T_c \sim 1\%$. In contrast, no hysteresis is encountered for the $n = 3$ model. Furthermore, the continuous variation of the order parameter for the $n = 3$ model constrains a fit to a simple power law, $\Phi(t) \sim (-t)^\beta$ for $0.01 \lesssim -t = (T_c - T)/T_c \lesssim 0.25$, with a critical exponent value of $\beta = 0.34 \pm 0.04$. This value is consistent with critical behavior of the

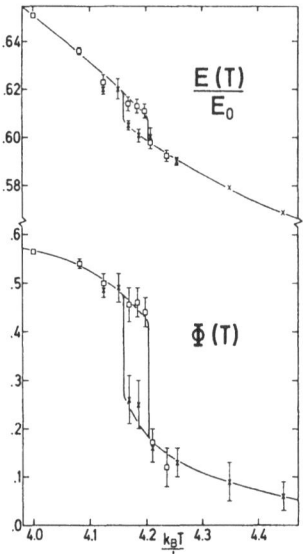

Fig. 4.1.3. Temperature dependence of the normalized internal energy and the order parameter in the transition region of the $n = 6$ model. Crosses and squares indicate Monte Carlo data obtained from decreasing and increasing temperature scans, respectively. The lattice has 14^3 sites.

$n = 3$ $d = 3$ Heisenberg universality class ($\beta = 0.3645 \pm 0.0025$, Le Guillou and Zinn-Justin 1980) and with the experimental result ($\beta = 0.32 \pm 0.02$) for $RbMnF_3$ (Tucciarone et al. 1971).

One of the striking similarities between the Monte Carlo order parameters of the two models in Fig. 4.1.2 is their pronounced high-temperature tails. These tails are caused by finite-size effects and are consequences of the persistence of long-range correlations in the vicinity of the phase transitions. Obviously, the $n = 6$ model supports critical-like fluctuations close to the transition point. The fluctuations are as pronounced as in the three-dimensional Ising model, cf. Fig. 3.1.2, and markedly different from those found in models which undergo ordinary first-order phase transitions, cf. Fig. 3.2.2. All this evidence supports the renormalization group prediction that the first-order transition in the $n = 6$ model is induced by fluctuations (cf. also Sec. 5.3.4).

As pointed out above, the $n = 6$ model should be an appropriate model for the magnetic phase transitions of UO_2 and $NdSn_3$. Neutron diffraction experiments on UO_2 have indeed shown the transition in this material to be of first order (Frazer et al. 1965) in agreement with our theoretical calculations. Recently, Hastings et al. (1980) studied $NdSn_3$ by neutron diffraction and found the order parameter to vary continuously through the transition region with no measurable hysteresis or critical scattering. Thus the transition appears neither as a distinct first-order nor as a continuous transition. Hastings et al. concluded that the transition is »unusual« in that the long-range order persists above the apparent transition temperature. It cannot be completely ruled out that random inhomogeneities may cause this unusual behavior. Thus, so far there is no clear-cut experimental determination of the nature of the magnetic transition in $NdSn_3$.

4.1.4 Crossover from first-order to continuous transitions in a symmetry-breaking field

We now revert to the question posed in Sec. 4.1.1: Do symmetry-breaking fields restore continuous transitions in systems with fluctuation-induced first-order transitions? The question has been approached in a paper by Knak Jensen et al. (1979) who carried out a combined study of Monte Carlo simulations, high-temperature series analyses, and renormalization group calculations on the $n = 6$ model presented in Sec. 4.1.3.

The symmetry-breaking field is chosen as a uniaxial stress modelled by an appropriate anisotropy in the coupling distributions J_1 and K of Eq. (4.1.1). With a stress parameter σ, the couplings take the form

$$J_1(\bar{r}_{jk}) = J_1(1 + \sigma), \quad K(\bar{r}_{jk}) = K(1 + \sigma), \quad \bar{r}_{jk} \parallel [001] \tag{4.1.5}$$

$$J_1(\bar{r}_{jk}) = J_1, \quad K(\bar{r}_{jk}) = K, \quad \bar{r}_{jk} \perp [001]. \tag{4.1.6}$$

The stress singles out the [001]-direction and lowers the space group symmetry from the cubic group $Pm3m$ to the tetragonal group $P4/mmm$. The $n = 6$ representation corresponding to the order parameter splits into three $n = 2$ representations, and the phase transition is therefore described by an $n = 2$ Landau-Ginzburg-Wilson Hamiltonian. This Hamiltonian has cubic anisotropy in contrast to the $n = 2$ Hamiltonian describing MnO in a uniaxial [111]-stress. A positive σ favors the order parameters Φ_{1z} and Φ_{2z}, whereas a small negative stress, $\sigma > -2$, stabilizes the components Φ_{3x} and Φ_{3y}. The stable structure changes for $\sigma < -2$ at low temperatures into an antiferromagnetic structure, which is not among the six structures degenerate for $\sigma = 0$. This structure is described by a propagation vector $\bar{k} = \frac{2\pi}{r_o}(0, 0, \frac{1}{2})$ with sublattice magnetization parallel to $\bar{k}$, that is a system classified by $n = 1$. This structure labeled Φ_{3z} is also shown in Fig. 4.1.1.

The renormalization group predictions for the $n = 6$ model in a symmetry-breaking uniaxial stress are obtained by mapping the lattice Hamiltonian, Eqs. (4.1.1), onto a continuous Landau-Ginzburg-Wilson Hamiltonian (Knak Jensen et al. 1979). The associated renormalization group flow diagram, to order ϵ, is shown in Fig. 4.1.4 in the case of negative stress. The diagram, spanned by the two renormalized fourth-order couplings u_1' and u_2', is divided into two regions by the line $u_2' = 6u_1'$. For $u_2' > 6u_1'$, the flow is towards an unstable region with no stable fixed points, i.e. first-order transitions. For $u_2' < 6u_1'$, the flow is towards the isotropic $n = 2$ fixed point. Along the line $u_2' = 6u_1'$, the model is tricritical (Rudnick 1978). As a function of σ, the initial Hamiltonian moves along the dashed line in Fig. 4.1.4. The calculation shows that the tricritical value of the stress is $\sigma_t = -0.025$. In view of the approximations involved in the renormalization group calculation, the precise value of σ_t should not be ascribed much importance. The crucial point is that for the $n = 6$ model, renormalization group theory predicts

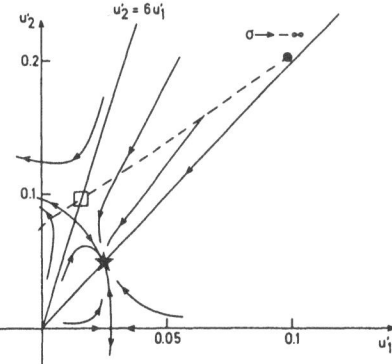

Fig. 4.1.4. Renormalization group flow diagram for the $n = 6$ model in a uniaxial stress, $\sigma < 0$, which stabilizes an $n = 2$ representation (Φ_{3x}, Φ_{3y}). The domain of attraction to the stable fixed point (asterix) is between the lines $u_2' = 6u_1'$ and $u_2' = 0$ in the first quadrant. The dashed line shows the starting Hamiltonian versus stress. For infinite (negative) stress, the Hamiltonian is indicated by a dot, which is within the stable regime. When the absolute value of the stress is lowered, the Hamiltonian moves towards the unstable regime, and at the tricritical value, $\sigma_t = -0.025$, it crosses the line $u_2' = 6u_1'$ at the point indicated by a square. For smaller absolute values of the stress, the Hamiltonian is unstable and the transition is of first order.

that a *finite* (although small) value of the symmetry-breaking stress is required to destroy the first-order phase transition. At positive stress, the calculation shows that the first-order transition survives for all finite values of σ.

In order to test these phenomenological renormalization group predictions of fluctuation-induced tricritical behavior, the complete phase diagram of the three-dimensional microscopic model has been mapped out by computer simulation techniques and by high-temperature series analysis.

The Monte Carlo calculations are performed on lattices with $N = 14^3$, 16^3, and 20^3 spins. The strategy employed is to choose progressively larger lattice sizes as the transition points are approached. It is particularly important to use a number of different lattice sizes in order to locate a tricritical point. Again, the distribution functions of the various order parameter components are indispensable in order to determine the equilibrium value of the long-range order. The usefulness of these distribution functions is discussed in Sec. 2.2.6. The calculations are performed for a series of positive and negative values of σ. In order to locate phase boundaries between ordered and disordered regions, as well as boundaries between different ordered phases, the phase diagram is scanned both along paths of constant stress and constant temperature.

Figures 4.1.5 and 4.1.6 show the Monte Carlo results for the order parameter as a function of temperature. By subjecting these result (and the corresponding results for the internal energy) to the criteria given in Sec. 2.2.9 for distinguishing between first-order and continuous transitions, we arrive at the following conclusion: *The fluctuation-induced first-order phase transition survives for small negative values of the symmetry-breaking stress.* Somewhere between $\sigma = -0.05$ and -0.30 there is a tricritical point and a crossover to continuous transitions. For positive stress, the

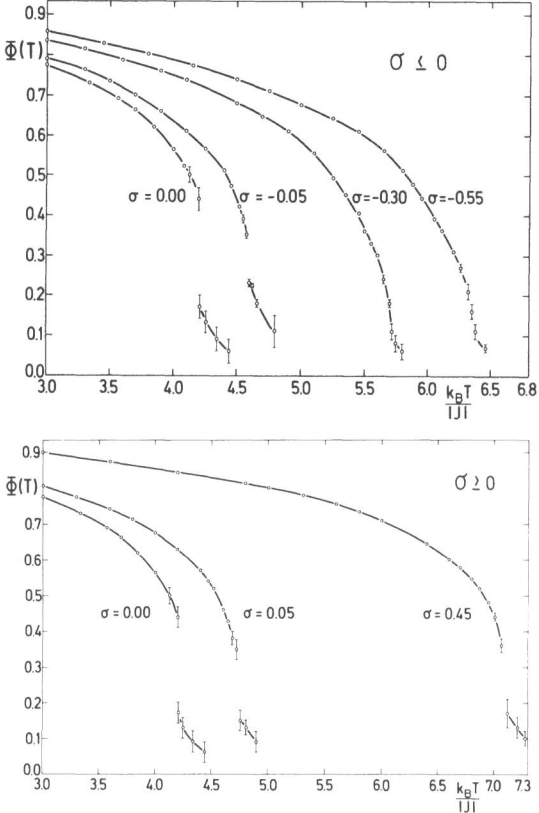

Fig. 4.1.5. Temperature dependence of the order parameter for the $n = 6$ model in a uniaxial stress, $\sigma \leq 0$.

Fig. 4.1.6. Temperature dependence of the order parameter for the $n = 6$ model in a uniaxial stress, $\sigma \geq 0$.

first-order transition persists up to the largest stress value investigated, $\sigma = 0.45$. Since the discontinuities decrease with increasing σ, it is likely that the characteristics of first-order transitions will disappear when σ gets sufficiently large. These conclusions are in close agreement with the renormalization group predictions and strengthen the concept of fluctuation-induced tricritical phenomena.

The high-temperature series expansions for the $n = 6$ model in a symmetry-breaking stress are derived according to the principles described in Sec. 2.3 (Mouritsen 1979). From the transverse and longitudinal correlation function series, the ordering susceptibility series (to sixth order in the inverse temperature) are constructed for the three different kinds of type-III antiferromagnetic ordering, cf. Fig. 4.1.1. The series coefficients appear as polynomials in $(1 + \sigma)$. The series have been subject to standard analysis including ratio and log-Padé analyses in terms of simple power-law singularities. In order to remove non-physical singularities and to improve convergence, various Euler transformations have been applied to the series. The phase boundary and the corresponding type of ordering, which is the stable one immediately below the transition point, are determined from the susceptibility series which has its physical singularity at the highest temperature. The results of the series analyses for locating the phase boundaries are as follows: The convergence is good for $-2 \leq \sigma \leq -0.30$ but becomes progressively poorer when σ is increased. For $\sigma \gtrsim -0.25$, it is not possible to extract unambiguous information from the Padé analyses. For $\sigma \gtrsim 0.45$, the convergence is smooth, although the series are not as well-behaved as for $\sigma \leq -0.30$. Decreasing σ from 0.45 to 0.35 leads to increasingly poor convergence and for $\sigma \lesssim 0.35$ no unambiguous information can be extracted from asymptotic analysis of the series. Finally, in the region $\sigma \leq -2$, the series is smooth and well behaved.

In Fig. 4.1.7 is shown the complete diagram in the (T, σ)-plane for the $n = 6$ model in a symmetry-breaking uniaxial stress. The figure encompasses Monte Carlo as well as series results. The series analysis based on an assumed power-law singularity gives no clear information on the location of the phase boundary for $-0.30 \lesssim \sigma \lesssim 0.40$. This may conceivably be a finite-series effect or it may be attributed to some physical mechanism. That the latter possibility is the more likely explanation is suggested by a similar series analysis of the $n = 3$ model in a uniaxial symmetry-breaking stress (Mouritsen and Knak Jensen 1980b). It turns out that the series for this model are smooth and easy to analyze in terms of a power-law singularity on both sides of $\sigma = 0$. The resulting phase diagram for the $n = 3$ model is shown in Fig. 4.1.8. Consequently, we find it justified to make a physical interpretation of the failure of the series analysis for the $n = 6$ model as an indication of an *effective crossover from a continuous to a first-*

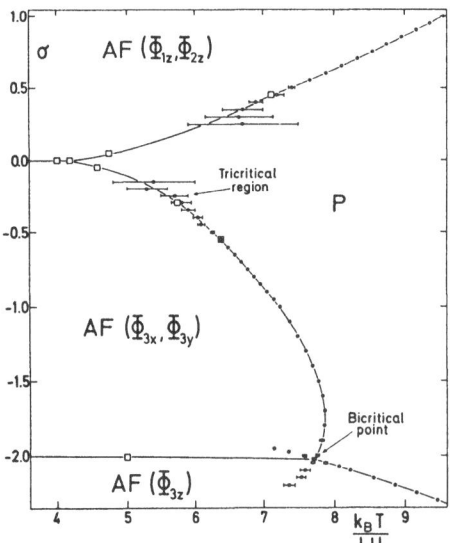

Fig. 4.1.7. Phase diagram of the $n = 6$ model in a symmetry-breaking uniaxial stress, σ. Circles indicate results from analysis of high-temperature series for the ordering susceptibility. Squares indicate Monte Carlo results. P denotes the paramagnetic phase. The spin structures corresponding to the three antiferromagnetic phases (*AF*) are shown in Fig. 4.1.1.

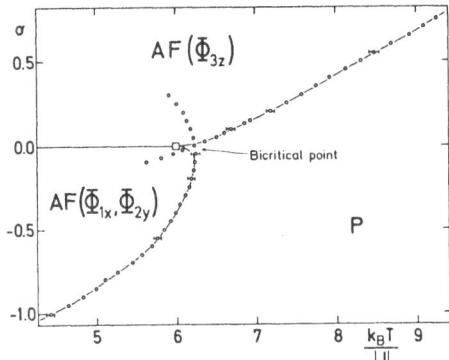

Fig. 4.1.8. Phase diagram of the $n = 3$ model in a symmetry-breaking uniaxial stres, σ. Circles indicate results from high-temperature series analysis. The square is a Monte Carlo result. The spin structures corresponding to the two antiferromagnetic phases (AF) are shown in Fig. 4.1.1. P denotes the paramagnetic phase. The uniaxial stress breaks the cubic symmetry and splits the $n = 3$ representation of the order parameter at $\sigma = 0$ into an $n = 2$ representation and an $n = 1$ representation. A positive σ favors the component Φ_{3z}, and a small negative σ favors the components Φ_{1x} and Φ_{2y}.

order phase transition. For negative σ, we estimate the tricritical stress σ_t to be larger than $\sigma \approx -0.3$, which is consistent with the estimate from the Monte Carlo calculations. It is seen from Fig. 4.1.7 that this crossover to the *tricritical* region takes place over a finite region of stress values as expected in a finite series analysis.

The result of the series analysis is apparently at variance with the Monte Carlo results for $\sigma = 0.45$. The series analysis suggests a power-law singularity appearing at the same temperature as the first-order phase transition observed in the Monte Carlo experiments. In principle, no *finite* series for the ordering susceptibility can give a definite result concerning the order of the phase transition. The finite series reflects the behavior of the system some distance from criticality, and if the susceptibility in this region is large — as is indicated by fluctuations observed in the coarse grained averages of the order parameter in the Monte Carlo experiment — the series analysis is likely to lead to a continuous transition, irrespective of the actual order. Finally, these observations are in line with the renormalization group calculations predicting fluctuation-induced first-order transitions in this regime.

The phase boundaries above and below $\sigma = -2$ have been calculated beyond stability to determine the point of intersection. This point is a *bicritical point* which results from competition between two different ordering mechanisms (Kosterlitz et al. 1976). At the bicritical point, the two different antiferromagnetic phases, denoted $AF(\Phi_{3x}, \Phi_{3y})$ and $AF(\Phi_{3z})$, become critical simultaneously and three phase boundary lines meet.

The critical line separating the paramagnetic phase (P) from $AF(\Phi_{3z})$ corresponds to continuous transitions belonging to the Ising $(n = 1)$ universality class. The transition at the bicritical point is expected to be continuous with Heisenberg like $(n = 3)$ critical behavior. The series are well behaved at this point and the results of the extrapolation procedures are consistent with a continuous phase transition.

The last critical line entering the bicritical point corresponds for $\sigma \lesssim -0.30$ to continuous transitions which belong to the XY ($n = 2$) universality class. We observe a bulge in the boundary exposing the bicritical point like an umbilicus. The same characteristics of the phase boundaries are encountered in the $(T, H_{\parallel})$-plane of the phase diagram for the anisotropic Heisenberg antiferromagnet in a longitudinal homogeneous external field (Kosterlitz et al. 1976, Landau and Binder 1978, Mouritsen et al. 1980). The remaining part of the phase diagram consists of the first-order line $\sigma = 0$, which separates the two antiferromagnetic phases $AF(\Phi_{1z}, \Phi_{2z})$ and $AF(\Phi_{3x}, \Phi_{3y})$. The line meets two other first-order lines at a *triple point*.

The discussion of fluctuation-induced tricritical phenomena may be concluded by the following remarks: We have studied, by three independent approaches, the phase transitions for a three-dimensional microscopic model of an antiferromagnetic system. The symmetry of the magnetic structure causes the corresponding renormalization group equations to lack stable fixed points in $4 - \epsilon$ dimensions. Our calculations all confirm the assertion that the phase transition changes from first-order to continuous when a sufficiently large symmetry-breaking field is applied. This result is in marked contrast to the results of the mean-field theory which predicts a continuous transition for our model for all values of the symmetry-breaking field. The phase diagram derived from our calculations is thus essentially different from that given by mean-field theory. The various calculations demonstrate that it is important to be very careful when dealing with fluctuation-induced tricritical phenomena: *Although a small, finite symmetry-breaking field creates a stable fixed point, there is no guarantee that this fixed point will ever be reached.*

A tricritical point induced by a symmetry-breaking uniaxial stress has been observed in MnO by neutron scattering experiments (Bloch et al. 1975). That this tricritical point is indeed induced by the lowering of the order parameter dimensionality and not merely by magnetoelastic effects was convincingly demonstrated in an experiment on MnO subject to a hydrostatic (isotropic) pressure (Bloch et al. 1980). The transition remained of first order for all pressures investigated. The $n = 6$ model corresponds to the magnetic structure of UO_2. We suggest experiments on UO_2 designed to ascertain whether or not the phase transition changes to continuous when a uniaxial stress is applied along the cube edge. Since the parameters used in our calculations do not necessarily bear any resemblance to the actual parameters for UO_2, we are unable to estimate the tricritical field.

4.1.5 Fluctuation-induced first-order phase transitions in Ising models with competing interactions

So far, no violation has been found of the Bak-Krinsky-Mukamel criterion (Bak et al. 1976a) for a fluctuation-induced first-order phase transition. Experiments on a great variety of materials as well as different types of theoretical model calculations have beautifully strengthened the originally somewhat shaky hypothesis that the lack of stable fixed points in $4 - \epsilon$ dimensions leads to first-order phase transitions.

Thus it appears that the hypothesis is well founded in three dimensions and that the criterion may supplement the three classical and renowned Landau symmetry criteria concerning the order of phase transitions (Landau and Lifshitz 1969).

There is one instance where the experimental observation of a continuous transition has been interpreted as indicating a break-down of the above criterion. By neutron diffraction Ott et al. (1979) have studied the cerium monochalcogenides *CeSe* and *CeTe* which are $n = 4$ type-II face-centered cubic antiferromagnets with the spins oriented along the propagation vector. The magnetic structure is shown in Fig. 4.1.9. This magnetic structure leads to a Landau-Ginzburg-Wilson Hamiltonian which has no stable fixed point within the ϵ-expansion (Bak et al. 1976a). Therefore, a first-order transition should occur. On the contrary, Ott et al. concluded from their experiment that the magnetic transition in both *CeSe* and *CeTe* is continuous: There is no hysteresis, the magnetization curve is continuous (any discontinuity is less than 10%), and the effective critical exponents of the magnetization are the same for *CeSe* and *CeTe* and have a value (0.36) characteristic of three-dimensional critical behavior. Furthermore, no latent heat was found in thermal measurements (Hullinger et al. 1978). As discussed in detail in Sec. 2.2.9, it is impossible by any experiment to prove that a phase transition is continuous. A seemingly continuous transition may very well be of first-order associated with small (and unresolved) discontinuities and with critical-like fluctuations. Especially the latter quality is of importance in the present context where theory predicts the transition to be driven first order by order parameter fluctuations. In fact, it has been observed experimentally and by computer simulations (cf. Secs. 4.1.3 and 5.3.4) that some systems undergoing fluctuation-induced first-order phase transitions on many accounts display critical-like behavior. Thus, the experiments by Ott et al. do not necessarily indicate that the otherwise successful Bak-Krinsky-Mukamel criterion is incomplete.

A number of theoretical investigations have been reported which address the controversy outlined above. Using renormalization group techniques, Mukamel and Wallace (1979) studied two different *three*-dimensional models of $n = 4$ type-II face-centered cubic antiferromagnets to investigate whether the renormalization group arguments in $d = 4 - \epsilon$ dimensions can carry over to $d = 3$. The two models are respectively an $n = 4$ phenomenological Landau-Ginzburg-Wilson model and a three-dimensional spin - $\frac{1}{2}$ Ising model

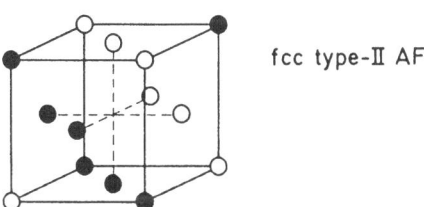

fcc type-II AF

Fig. 4.1.9. Magnetic structure of type-II face-centered cubic antiferromagnets. The two magnetic sublattices are indicated by white and black dots. The spins are either parallel or antiparallel to the [111]-direction. The structure is $n = 4$-fold degenerate corresponding to the four body diagonals of the cube.

$$H = J_1 \sum_{i,j}^{nn} \sigma_i \sigma_j - J_2 \sum_{i,j}^{nnn} \sigma_i \sigma_j \tag{4.1.7}$$

with $\sigma_i = \pm 1$, $J_1 > 0$, and $\alpha \equiv J_2/J_1 < -\frac{1}{2}$. The two summations are over nearest (nn) and next-nearest (nnn) neighbors. The Ising model in Eq. (4.1.7) was shown to have the same fourth-order invariants in its free energy expansion as has the phenomenological model. For both models, it was found that no fixed points become stable in three dimensions. Mukamel and Wallace suggested that the transitions observed in $CeSe$ and $CeTe$ should be interpreted as being »weakly first-order transitions.«

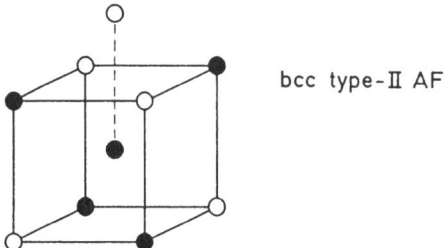

bcc type-Ⅱ AF

Fig. 4.1.10. Magnetic structure of type-II body-centered cubic antiferromagnets. The two magnetic sublattices are indicated by white and black dots.

Further support for fluctuation-induced first-order phase transitions in type-II face-centered cubic antiferromagnets comes from a Monte Carlo simulation study of the phase diagram of the Ising model in Eq. (4.1.7) with competing nearest and next-nearest neighbor interactions (Phani et al. 1979, 1980a). These authors calculated the complete phase diagram spanned by α and the temperature.[*] In the region of interest to us, $\alpha < -\frac{1}{2}$, where renormalization group calculations predict fluctuation-induced first-order transitions, the ground state is that shown in Fig. 4.1.9 and the order parameter has $n = 4$ components. The Monte Carlo results for a number of lattice sizes yield clear evidence of a first-order transition for $-1 \lesssim \alpha < -\frac{1}{2}$. Discontinuities are seen in the internal energy and in the order parameter, hysteresis and metastable states are encountered, and the approach towards equilibrium follows a characteristic two-step relaxation behavior in the transition region (cf. Sec. 2.2.9). However, for $\alpha \lesssim -1$ no indications of first-order transitions are found. It is impossible to decide whether this indicates a crossover from first-order to continuous transitions around $\alpha \sim -1$ or whether it is a simple consequence of decreasing discontinuities as α is lowered. Obviously, in the limit $\alpha \to -\infty$ a continuous transition will occur since in this limit the model decouples into four independent simple cubic Ising lattices.

A few other Monte Carlo studies of three-dimensional Ising models undergoing fluctuation-induced first-order transitions have been reported within the last few

[*] For a Monte Carlo study of the phase diagram of this model including a magnetic field, see Binder 1981a.

years. Among these are N-color Askin-Teller models (Ditzian et al. 1980, Grest and Widom 1981) and body-centered cubic Ising models with competing pair interactions (Banavar et al. 1979). The latter models constitute a particularly good example of how renormalization group methods, Monte Carlo simulations, and series analysis effectively complement one another in clarifying a difficult problem. Thus, we shall describe the results for this case in some detail.

The Ising model is again that of Eq. (4.1.7), this time arrayed on a body-centered cubic lattice. For $\alpha > \frac{2}{3}$, the magnetic ground state is a type-II body-centered cubic antiferromagnet as shown in Fig. 4.1.10. The order parameter has $n = 2$ components, corresponding to the two different ways that two interpenetrating type-I simple cubic antiferromagnetic structures can be embedded into a body-centered cubic lattice. Banavar et al. (1979) studied the Landau-Ginzburg-Wilson Hamiltonian appropriate for this Ising model and by analyzing the ϵ-expansion they found that the initial Hamiltonian is outside the range of attraction of any stable fixed points. Since mean-field theory for this model predicts the transition from the antiferromagnetic state to the paramagnetic state to be continuous, it was concluded that the transition is driven first-order by fluctuations. The conclusion was supported by computer simulations (Banavar et al. 1979). The Monte Carlo evidence for first-order transitions for $\alpha > \frac{2}{3}$ is very similar to that described above for the type-II face-centered cubic Ising model. The first order behavior is very pronounced for $\frac{2}{3} \lesssim \alpha \lesssim 0.7$. However, when α becomes sufficiently large, the transition of the finite lattices appears effectively as a continuous transition. Finally, an extensive series analysis study of the model has led to similar conclusions. By observing significant differences between transition temperatures obtained from matching high- and low-temperature free-energy series and transition temperatures obtained from analysis of ordering susceptibility series in terms of power-law singularities, Velgakis and Ferer (1983) established the transition to be of first order for $\frac{2}{3} \lesssim \alpha \lesssim 0.77$. For $\alpha \gtrsim 0.77$, the analysis is consistent with a continuous transition. Clearly, the Monte Carlo calculation and series analysis are in perfect accordance, although none of these numerical approaches are capable of determining whether or not the model has a tricritical point.

4.2 Critical Phenomena at Marginal Dimensionality

4.2.1 The role of a marginal spatial dimension

The renormalization group approach to critical phenomena makes a remarkable prediction: There exists a marginal spatial dimension, d^*, above which fluctuations become irrelevant to critical behavior (see e.g. Brézin et al. 1976). Above d^*, critical behavior is therefore expected to be exactly accounted for by mean-field (or Landau) theory. Below d^*, the critical behavior is non-trivial and renormalization group calculations lead to non-classical exponents as given by the asymptotic $\epsilon = (d^* - d)$-expansion (Wilson and Fisher 1972). Of particular interest is the case in which the marginal and the physical dimensions are the same, $d = d^*$. In this case, mean-field theory and fluctuation theory coalesce and the renormalization group equations can be solved exactly!

The situation $d = d^*$ was first considered by Larkin and Khmel'nitskii (1969) (see also Wegner and Riedel 1973, Brézin et al. 1973), who found that singular thermodynamic functions, $f(t)$, at the critical point are given by mean-field power laws modified by multiplicative logarithmic corrections

$$f(t) \sim |t|^q |\ln|t||^{p(n)}, \quad t = (T - T_c)/T_c \to 0. \tag{4.2.1}$$

q is the *mean-field* exponent for $f(t)$ and $p(n)$ is a universal exponent only depending on the number of components of the order parameter. More recently, corrections to the Larkin-Khmel'nitskii prediction have been derived (Brézin and Zinn-Justin 1976). These corrections, which become important away from the critical point, modify Eq. (4.2.1) according to

$$f(t) \sim |t|^q |\ln|t||^{p(n)} (1 + Q(n) \ln|\ln|t||/|\ln|t|| + O(|\ln|t||^{-1})). \tag{4.2.2}$$

The correction amplitude $Q(n)$ is universal in the sense that it does not depend on any details within the system but only on n and the nature of the interparticle interactions. In contrast, the additional correction terms are non-universal (Brézin and Zinn-Justin 1976). This makes experimental investigations of Eq. (4.2.2) a very delicate matter.

It is important to emphasize that the renormalization group predictions at marginal dimension, *viz* Eq. (4.2.2), do *not* rely on any approximative ϵ- or $1/n$-expansions. Systems with $d = d^*$ therefore constitute unique cases for performing a direct test of the whole renormalization group approach to critical phenomena. Now, the value of d^* for a particular system depends on the nature of the interparticle interactions. It turns out that there are only a few cases in which d^* coincides with the physical dimension. In fact, $d^* = 4$ for common short-range interaction systems. However, there is the particular case of long-range dipolar coupled uniaxial ferromagnets and ferroelectrics which has $d^* = 3$ (Larkin and Khmel'nitskii 1969). For this very reason, systems of the latter type have in recent years received considerable experimental attention. Among these are $LiTbF_4$ (Beauvillain et al. 1980 and references therein), $GdCl_3$ (Kötzler and Eiselt 1976), $Dy(C_2H_5SO_4)_3 \cdot 9H_2O$ (Frowein and Kötzler 1976), TbF_3 (Brinkmann et al. 1978), and $LiHoF_4$ (Griffin et al. 1980 and references therein). The experimental measurements on these systems of a variety of properties, such as magnetization, susceptibility, and specific heat, lend support to the renormalization group prediction of logarithmic corrections, Eq. (4.2.1). So far, corrections to the leading singularity as given by Eq. (4.2.2) have only been investigated experimentally for the magnetization of $LiHoF_4$ (Griffin et al. 1980), and the correction amplitudes are very poorly determined.

As already indicated, we are facing the very peculiar situation that the renormalization group prediction at marginal dimension may escape a critical *experimental* examination for the most common class of magnetic systems, i.e. three-dimensional systems with short-range exhange interactions. This is only the case, however, when the restriction is made to conventional experiments: *Computer experiments are readily performed in any integer dimension.*

4.2.2 Computer experiments on hypercubic Ising models: »A romance of many dimensions« *

It is fairly straightforward to devise a computer simulation program which works for arbitrary integer spatial dimensions. Monte Carlo simulations thus allow for independent »experimental« investigations of critical phenomena in unphysical marginal dimensions. This section reports the results of a comparative Monte Carlo study of simple hypercubic spin -$\frac{1}{2}$ short-range Ising ferromagnets (Mouritsen and Knak Jensen 1979a,b). To examine whether logarithmic corrections are unique to four dimensions for short-range interactions, the Monte Carlo simulations have been performed for $d = 3, 4$, and 5. The discussion will focus on the order parameter (spontaneous ferromagnetic magnetization), $\Phi(t)$. Results for the susceptibility, $\chi(t)$, and the critical isotherm, $\Phi(h, t = 0)$, will briefly be reported. h is the magnetic field conjugate to the order parameter. According to the renormalization group theory, the following versions of Eq. (4.2.1) are expected to hold for these functions

$$\Phi(t) \simeq B(-t)^{\frac{1}{2}}|\ln(-t)|^{\frac{1}{3}}, \quad t < 0 \tag{4.2.3}$$

$$\chi(t) \simeq \Gamma_-(-t)^{-1}|\ln(-t)|^{\frac{1}{3}}, \quad t < 0 \tag{4.2.4}$$

$$\Phi(h) \simeq Dh^{\frac{1}{3}}|\ln(h)|^{\frac{1}{3}}, \quad t = 0. \tag{4.2.5}$$

The computer simulations have been carried out for a number of lattice sizes for each value of d. The main calculations are performed on lattices with $N = 14^3, 12^4$, and 6^5 spins for $d = 3, 4$, and 5. The critical temperature, $T_c(N)$, of the finite lattice is obtained from the peak position of the susceptibility. We shall not here be concerned with a detailed analysis of the finite-size effects and the agreement between Monte Carlo critical temperatures and critical temperatures derived from series analysis (Mouritsen and Knak Jensen 1979a). For the present purpose, it suffices to state the result that finite-size effects are found to be unimportant for the lattice sizes mentioned above in $d = 4$ for $-t \gtrsim 0.005$ and in $d = 3$ and 5 for $-t \gtrsim 0.01$.

The Monte Carlo results for $\Phi(t)$ in $d = 3, 4$, and 5 dimensions are shown in Fig. 4.2.1 as a log-log plot. This figure also includes the exact result for $d = 2$ (Yang 1952). The critical temperatures needed for this plot are for $d = 3, 4$, and 5 those resulting from appropriate series analyses. The search for *leading* logarithmic corrections to the mean-field order parameter, cf. Eq. (4.2.3), proceeds in two steps. Firstly, we show that if the Monte Carlo data, $\Phi^{MC}(t)$, contain a factor $(-t)^\beta$ with the mean-field exponent $q = \beta = \frac{1}{2}$, it simultaneously has to include a multiplicative logarithmic correction. This is demonstrated in Fig.4.2.2 which gives a plot of the amplitude function

$$B(p, t) = \Phi^{MC}(t)/[(-t)^{\frac{1}{2}}|\ln(-t)|^p] \tag{4.2.6}$$

---

* Abbott 1874.

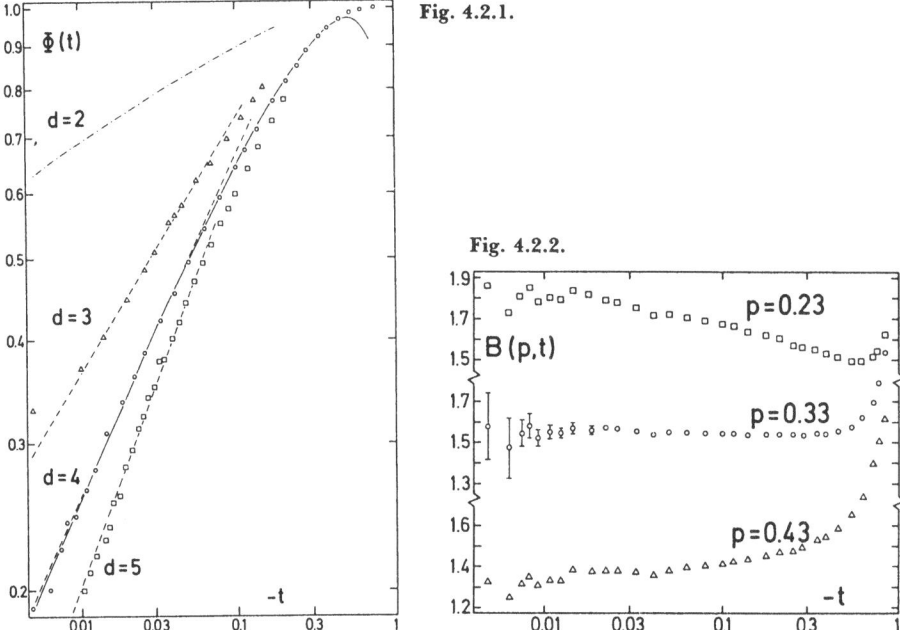

Fig. 4.2.1.

Fig. 4.2.2.

Fig. 4.2.1. Log-log plot of the order parameter $\Phi(t)$ vs $-t = (T_c - T)/T_c$ for d-dimensional Ising models. Dot-dashed line signifies exact results for $d = 2$. Triangles, circles, and squares denote Monte Carlo data for $d = 3, 4$, and 5. The solid line is the function $\Phi(t) = 1.545(-t)^{\frac{1}{2}}|\ln(-t)|^{\frac{1}{3}}$. The dashed lines represent simple power laws $\Phi(t) = B(-t)^{\beta}$ with parameters B and β as given in the text for the various values of d.

Fig. 4.2.2. Analysis of Monte Carlo data for the order parameter of the $d = 4$ Ising model. $B(p, t)$ is defined in Eq. (4.2.6).

for various values of p. The figure clearly establishes the presence of a logarithmic correction, $|\ln|t||^{0.33}$, over about two decades. Secondly, a full three-parameter fit of Eq. (4.2.1) to $\Phi^{MC}(t)$ is performed by means of a non-linear least-squares fitting procedure. The fitting is carried out for different ranges, $-t \in [0.005, t_{max}]$, of reduced temperatures. The quality of the fit is extremely good and the fit is very sensitive to the three parameters. For values of t_{max} up to about 0.4, the fit is stable in the three parameters with $B = 1.545, p = \frac{1}{2}$, and $q = \frac{1}{3}$. For $t_{max} \lesssim 0.06$, an equally good fit is obtained by the effective power law

$$\Phi(t) \simeq 1.73(-t)^{0.41}. \tag{4.2.7}$$

Consequently, our strongest piece of evidence for the presence of logarithmic corrections derives from the data above $-t \sim 0.06$ where the data are more accurate and the logarithmic factor is more dominant. Figure 4.2.1 includes the final fitting functions with and without the logarithmic correction.

The same methods of analysis have been applied to the Monte Carlo order parameter data for $d = 3$ and $d = 5$. In both cases, no evidence is found for a logarithmic correction factor to a pure power law, $\Phi(t) = B(-t)^\beta$. The results of the fitting are

$$d = 3: \quad B = 1.5 \pm 0.1; \quad 0.290 < \beta < 0.315, \quad 0.01 < -t < 0.10 \tag{4.2.8}$$

$$d = 5: \quad B = 2.0 \pm 0.2; \quad 0.470 < \beta < 0.530, \quad 0.01 < -t < 0.06. \tag{4.2.9}$$

For $d = 3$, the Monte Carlo result is consistent with series analysis. In particular, the value of β is non-classical. On the contrary, the value of β for $d = 5$ is consistent with the mean-field value $\beta = \frac{1}{2}$. The power laws with the parameters in Eqs. (4.2.8) and (4.2.9) are also shown in Fig. 4.2.1. Our finding of mean-field exponents in $d = 4$ has been confirmed by a recent Monte Carlo renormalization group study (Blöte and Swendsen 1980) of the four-dimensional Ising model. In that study, the critical behavior is investigated *at* the critical point, $t = 0$, and no information about correction terms is available.

It is interesting to note that the value of the effective exponent, $\beta_{eff} \simeq 0.41$, found from a fit to a pure power law, Eq. (4.2.7), for the $d = 4$ Ising model is similar to the effective exponents found experimentally for a series of uniaxial dipolar ferromagnets. Specifically, for TbF_3, $Dy(C_2H_5SO_4)_3 \cdot 9H_2O$, $GdCl_3$, and $LiTbF_4$ β_{eff} varies from 0.39 to 0.45. These anomalous values of the order parameter exponent are consistent with the renormalization group prediction, Eq. (4.2.3), considering that $|\ln|t||^{\frac{1}{3}}$ looks very much like the power law $|t|^{-0.07}$ in the reduced temperature interval under consideration.

In conclusion, we have demonstrated that the Monte Carlo calculations provide strong experimental evidence for a leading multiplicative logarithmic correction to the mean-field order parameter of the four-dimensional Ising model. Furthermore, logarithmic corrections are found to be unique to four dimensions. The form of the logarithmic correction agrees with that predicted by renormalization group theory.

We now proceed to look for corrections to the leading singularity of the order parameter for the $d = 4$ Ising model (Mouritsen and Knak Jensen 1979b). This is a much more delicate matter since a fit to experimental data involves a substantial number of parameters, cf. Eq. (4.2.2). Furthermore, the subleading terms vary extremely slowly with temperature. Here we shall assume the values of the exponents q and $p(n)$ as given by the renormalization group theory and then examine whether the Monte Carlo data can constrain a fit to the form

$$\Phi(t) = B|t|^{\frac{1}{2}}|\ln|t||^{\frac{1}{3}}(1 + Q\ln|\ln|t||/|\ln|t||). \tag{4.2.10}$$

Such an examination requires high-precision Monte Carlo data for $\Phi(t)$ in the region $-t > 0.4$ where the form in Eq. (4.2.3) does not fit the data (cf. Fig. 4.2.1). Again, a non-linear least-squares fitting to the data is carried out, this time in terms of the two parameters B and Q. For $-t \lesssim 0.56$, the fit is stable in the parameters and very sensitive to the values of B and Q. The resulting values are

$$B = 1.543 \pm 0.02, \quad Q = -0.030 \pm 0.02. \tag{4.2.11}$$

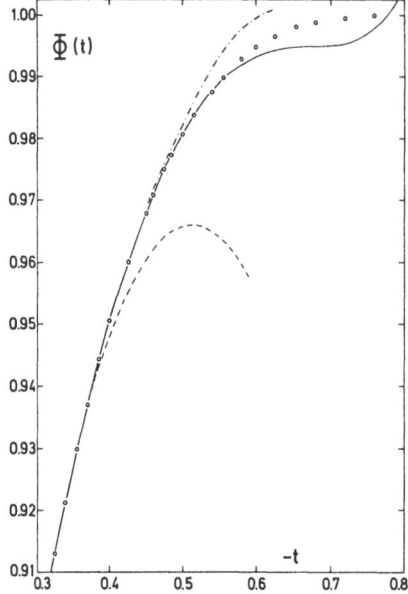

Fig. 4.2.3. Order parameter $\Phi(t)$ vs reduced temperature for the $d = 4$ Ising model. Circles: Monte Carlo data; full line: $\Phi(t) = 1.543(-t)^{\frac{1}{2}}|\ln(-t)|^{\frac{1}{3}}[1 - 0.030 \ln|\ln(-t)|/|\ln(-t)|]$; dashed line: $\Phi(t) = 1.543(-t)^{\frac{1}{2}}|\ln(-t)|^{\frac{1}{3}}$; dot-dashed line: Kogon's fit including a crossover scaling function.

The corresponding fitting function is shown in Fig. 4.2.3. The value of B is consistent with that obtained by neglecting the subleading corrections. Subsequently, Kogon (1981) has calculated the renormalization group value of Q to be $Q = -\frac{25}{81}$, i.e. the Monte Carlo result is an order of magnitude too low! In a first attempt to resolve this discrepancy, we have included a second subleading correction term in Eq. (4.2.10), $P/|\ln|t||$, in order to investigate whether Q effectively accounts for such a term. This is not the case, however, since the best fit is obtained for $P \simeq 0$. In a second attempt, Kogon (1981) has analyzed the Monte Carlo data via the form in Eq. (4.2.10) by fixing $Q = -\frac{25}{81}$ and by replacing t by t/t_o, where t_o ($\gg t$) is a fitting parameter which, according to the original suggestion made by Aharony and Halperin (1975), may effectively describe crossover between different types of critical behavior. Kogon finds that a two-parameter fit in terms of t_o and B is very poor and strongly dependent on t_{max}. Finally, Kogon has used a complete crossover scaling function (Kogon 1981) in a fit to the data. This function describes the crossover from Ising to Gaussian behavior as the temperature is lowered (Bruce and Wallace 1976). The resulting fit, as shown in Fig. 4.2.3, successfully describes the Monte Carlo data up to $-t \simeq 0.46$. Although the final fit reproduces the order parameter data over a smaller range of reduced temperatures, it is a considerable improvement over the previous fits of the asymptotic forms since the final fitting function incorporates the correct value of the correction amplitude Q.

4.2.3 Susceptibility and critical isotherm of the four-dimensional Ising model

The Monte Carlo results for the susceptibility, * $\chi(t)$, of the $d = 4$ Ising model for $T < T_c$ are much less accurate than are the corresponding order parameter data. Therefore, the data for $\chi(t)$ cannot sustain a detailed analysis in terms of logarithmic corrections. However, as shown in Fig. 4.2.4, the Monte Carlo results are certainly consistent with logarithmic corrections as given by Eq. (4.2.4) over a substantial range of reduced temperatures, $0.005 \lesssim -t \lesssim 0.12$. Since $|\ln(-t)|^{\frac{1}{3}}$ varies very slowly in this temperature region, the error bars in Fig. 4.2.4 also allow a fit of a pure power law, $\chi(t) \sim (-t)^{-\gamma_{eff}}$, with $\gamma_{eff} \simeq 1.13$. We are unable to prefer one of these fits to the other. The same predicament recurs when analyzing the experimental susceptibility of uniaxial dipolar ferromagnets where simple power laws cannot be completely ruled out. However, by insisting on simple power-law behavior, one obtains anomalous values of the exponent γ_{eff} which typically lie in the range from 1.05 to 1.15 for the uniaxial dipolar ferromagnets. The ratio between the amplitude, Γ_-, of the susceptibility for $T < T_c$ and the corresponding amplitude, Γ_+, for $T > T_c$ is predicted to be a universal number, $\Gamma_+/\Gamma_- = 2$ (Aharony and Halperin 1975). Γ_+ has been derived from an analysis of the high-temperature series expansion of the susceptibility in terms of the renormalization group prediction, Eq. (4.2.4) (McKenzie and Gaunt 1980). The result, $\Gamma_+ \simeq 0.068$, in combination with the Monte Carlo estimate, $\Gamma_- \simeq 0.023$, from Fig. 4.2.4, leads to $\Gamma_+/\Gamma_- \simeq 3$. In view of the uncertainty in estimating Γ_- from Fig. 4.2.4, we do not consider this as a serious discrepancy. The amplitudes

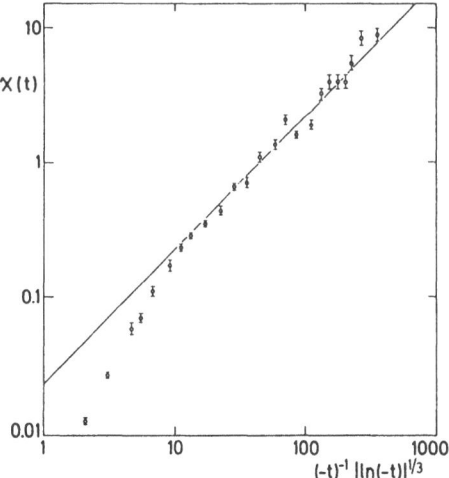

Fig.4.2.4. Log-log plot of the susceptibility vs $|t|^{-1}|\ln|t||^{\frac{1}{3}}$ for the $d = 4$ Ising model. The Monte Carlo data (circles) are obtained for a system with $N = 12^4$ spins. The solid line (with unit slope) is the renormalization group prediction, Eq. (4.2.4), with $\Gamma_- \simeq 0.023$.

---

* O.G. Mouritsen, unpublished calculations.

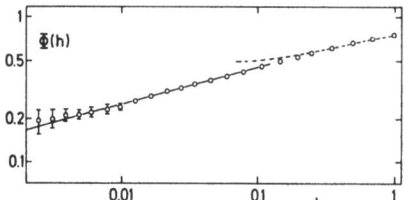

Fig. 4.2.5. Log-log plot of the critical isotherm, $\Phi(h, t = 0)$, of the $d = 4$ Ising model. Monte Carlo data (circles) are given for a system with $N = 12^4$ spins. The dashed line is the bare low-temperature series expansion to order $[\exp(-2h/k_B T)]^{15}$ (McKenzie et al. 1979). The solid line is the effective power law, Eq. (4.2.12), with $\delta_{eff} = 4$.

extracted from the Monte Carlo data as well as from the series expansion may be affected by neglecting subleading correction terms.

Finally, we present some preliminary Monte Carlo results for the critical isotherm, $\Phi(h)$, of the $d = 4$ Ising model.* The data are obtained for a large Ising lattice with $N = 12^4 = 20736$ spins subjected to a magnetic field, h. The calculations are performed precisely at the critical temperature, T_c. T_c is obtained from an asymptotic analysis of the high-temperature susceptibility series to order $[\tanh(J/k_B T)]^{17}$ in terms of a logarithmically corrected mean-field singularity (Gaunt et al. 1979). The Monte Carlo data are presented in Fig. 4.2.5. Within the accuracy, $\Phi(h)$ does not display sufficient structure to allow a direct analysis in terms of logarithmic corrections. In fact, for $h \lesssim 0.2$ the Monte Carlo data in Fig. 4.2.5 are effectively described by a simple power law

$$\Phi(h) \sim h^{q_{eff}}, \quad q_{eff} = 1/\delta_{eff} \tag{4.2.12}$$

with an effective exponent $1/\delta_{eff} \simeq 0.25$. Noting again that $|\ln h|^{\frac{1}{3}} \simeq h^{-0.07}$ for the range of field values considered here, we conclude that the Monte Carlo data for the critical isotherm are consistent with the renormalization group prediction, Eq. (4.2.5), with a mean-field exponent, $q^{-1} = \delta = 3$.

4.2.4 Conclusions on critical behavior in marginal dimensions

The combined evidence obtained from experiments on uniaxial ferromagnets as well as from series expansions and computer studies on the four-dimensional Ising model suggests that the renormalization group prediction, Eq. (4.2.1), of logarithmic corrections in marginal dimensions is well established. The fitting of the *leading* singularity to experimental and numerical data for the order parameter appears to be reliable and unambiguous.

The lesson taught from the fitting of experimental and numerical data to functions including *subleading* logarithmic corrections, cf. Eq. (4.2.2), is that extreme care should be taken when the fits are extended to cover regions far from the

---

* O.G. Mouritsen, unpublished calculations.

critical point. With the present quality of experimental data, the tremendous flexibility afforded by forms like Eq. (4.2.2) and the fact that crossover between different kinds of critical behavior may occur, preclude a critical examination of the renormalization group predictions of subleading logarithmic corrections to critical behavior in marginal dimensions. This situation closely resembles what has recently been described for experimental tests of renormalization group predictions concerning correction-to-scaling at ordinary critical points (Güttinger and Cannell 1981, Balzarini and Mouritsen 1983).

In closing our »romance of many dimensions«, we would mention that four-dimensional Ising lattice models may indeed be less exotic and physically more relevant than what seems immediately apparent. Ising models and their phase structure in four space-time dimensions have proved extremely useful to clarify aspects of lattice gauge theories which at present seem promising for demonstrating hadronic quark confinement in elementary particle physics (for a review on Monte Carlo simulations of lattice gauge theories, see e.g. Rebbi 1980).

4.3 Basic Assumptions of Critical Correlation Theories

4.3.1   Review of a critical correlation theory

The search for a solution to the three-dimensional Ising model has generated a number of approximate analytical calculation schemes to derive properties of the phase transition. Many of these schemes are based on assumptions which are uncontrolled. This is also the case in the critical correlation theory originally advanced by Frank and Mitran (1977) to determine critical temperatures of the ferromagnetic spin$-\frac{1}{2}$ Ising model on the cubic lattices. Since then, the critical correlation theory has been applied to a series of different Ising problems (for a list of references describing the development of the theory and its applications, see e.g. Frank et al. 1982, Frank and Cheung 1984).[*] Despite later modifications, the central core of the theory remains the same. This core involves a basic assumption about the behavior of certain correlation functions at the critical point. It is the question of the validity of this assumption that we want to address now.

The starting point of the theory is the exact correlation identity

$$< \sigma_i F > = \frac{1}{2} < F \tanh[\frac{1}{2} K(O_i + h_i)] >, \qquad (4.3.1)$$

where F is an arbitrary spin operator which is a combination of sums of products of spin operators that do not include σ_i (Suzuki 1965, Callen 1963). h_i is a site-dependent external magnetic field,

$$O_i = \sum_{j=1}^{z} J_{ij}\sigma_j, \quad J_{ij} = J > 0, \qquad (4.3.2)$$

---

[*] For related theories, see Taggart and Fittipaldi 1982, Zhang 1981.

is the local field created by the z nearest neighbors interacting with i, and $K = 1/k_B T$. In the limit $F = 1$, Eq. (4.3.1) yields the exact expression for the Ising ferromagnetic order parameter

$$< \sigma_i > = \tfrac{1}{2} < \tanh[(\tfrac{1}{2} K (O_i + h_i)] > . \tag{4.3.3}$$

From the exact Eq. (4.3.1), the following identities follow by putting $h_i = 0$ (all i) and, successively, $F = \sigma_j (j \neq i)$, $\tanh(\tfrac{1}{2} K O_i)$, and O_i

$$< \sigma_i \sigma_j > = \tfrac{1}{2} < \sigma_j \tanh(\tfrac{1}{2} K O_i) >, \quad i \neq j$$

$$= \tfrac{1}{4}, \qquad i = j \tag{4.3.4}$$

$$< \sigma_i \tanh(\tfrac{1}{2} K O_i) > = \tfrac{1}{2} < \tanh^2(\tfrac{1}{2} K O_i) > \tag{4.3.5}$$

$$< \sigma_i O_i > = \tfrac{1}{2} < O_i \tanh(\tfrac{1}{2} K O_i) > . \tag{4.3.6}$$

The theory now proceeds towards a determination of the critical temperature T_c by introducing an *ad hoc* basic assumption (Frank and Mitran 1977). Observing that, in the limits $T \to T_c$ and $h_i \to 0$ (all i), the order parameter in Eq. (4.3.3) vanishes and that therefore also every term, $< O_i^{2n+1} >$, of the power series expansion of the right-hand side of Eq. (4.3.3) has to vanish, Frank and Mitran make the assumption that $< O_i^{2n+1} >$ ($n = 1, 2, \ldots$) and $< O_i >$ approach zero in the same way, i.e.

$$< O_i^{2n+1} > = F_{2n} < O_i > . \tag{4.3.7}$$

F_{2n} is a set of wellbehaved analytical functions of the h_i. Differentiation of Eq. (4.3.7) with respect to h_j yields

$$< \sigma_j O_i^{2n+1} > = F_{2n} < \sigma_j O_i > + < O_i > K^{-1} \frac{d}{dh_j} F_{2n}. \tag{4.3.8}$$

Since $< O_i >$ vanishes for $T \to T_c$ whereas the pair correlation $< \sigma_j O_i >$ does not, the basic assumption* is seen to imply

$$< \sigma_j O_i^{2n+1} > = F_{2n} < \sigma_j O_i > \tag{4.3.9}$$

where F_{2n} is independent of j. F_{2n} is determined by self-consistency, e.g. by multiplying Eq. (4.3.9) by J_{ij} and summing over j yielding $F_{2n} = < O_i^{2n+2} > / < O_i^2 >$ (Frank et al. 1982).

A site-independent quantity, A, is then defined by

$$\frac{2A}{zJ} \equiv \frac{< \sigma_j \tan(\tfrac{1}{2} K O_i) >}{< \sigma_j O_i >}. \tag{4.3.10}$$

---

* Other versions of the basic assumption are considered in Frank et al. 1982, Frank and Cheung 1984.

Applying Eqs. (4.3.5) and (4.3.6), Eq. (4.3.10) becomes

$$\frac{2A}{zJ} = \frac{< \tanh^2(\frac{1}{2}KO_i) >}{< O_i \tanh(\frac{1}{2}KO_i) >}.$$ (4.3.11)

We are now in a position to derive a criticality condition from which T_c may be obtained. By taking the Fourier transform of the pair correlation function, Eq. (4.3.4), with Eq. (4.3.10) inserted, we obtain the ordering susceptibility to be

$$\chi(\overline{q}) = C[1 - AJ(\overline{q})/zJ]^{-1}.$$ (4.3.12)

C is a function of lattice structure and $J(\overline{q})$ is the Fourier transform of the coupling distribution, J_{ij}. At the ferromagnetic transition, $\overline{q} = \overline{0}$, $\chi(\overline{q})$ diverges, and the condition for criticality is simply given by

$$A = 1, \quad T = T_c.$$ (4.3.13)

The criticality condition, Eq. (4.3.13), imposes a number of restrictions on the critical correlation functions from which the critical temperature may eventually be derived. Since knowledge of that part of the theory is not required for our present purpose, it suffices to refer the reader to Frank et al. (1982) and Frank and Cheung (1984) for a complete description. The important point to stress is that the critical correlation theory based on the assumption, Eq. (4.3.9), has led to amazingly accurate values for the critical temperatures of the three cubic lattices (Frank and Mitran 1977, Frank et al. 1982, Frank and Cheung 1984).[*] Depending essentially on the version of the basic assumption used and on the lattice structure, critical temperatures have been derived which are within 0.15 - 2% of the estimates from series expansions. This is indeed a remarkably successful result for an analytical theory of the three-dimensional Ising problem.

4.3.2   Testing the basic assumption by Monte Carlo calculations

The physical substance of the basic assumption of the critical correlation theory, Eq. (4.3.7), remains rather unclear. Certainly, the favorable comparison with series expansions indicates that the assumption is useful. But is it correct, and if not, is it a good approximation? In this section, we shall describe a *direct* numerical investigation of the validity of the basic assumption.

The investigation is based on a Monte Carlo calculation of selected critical correlation functions for the spin $-\frac{1}{2}$ Ising model on the simple cubic and diamond lattices (Frank et al. 1982, Mouritsen 1980). Simulation techniques seem superior for this purpose since the multi-spin correlation functions entering the higher-order versions of Eq. (4.3.7) would be very difficult to determine by other methods, e.g. by series expansions. We have chosen to test the following relations:

---

[*] For an extension to $d > 3$ hypercubic lattices, see Nath and Frank 1982.

$$P_1 = \frac{<O_i^2>}{zJ<\sigma_i O_i>} \quad (= 1, \quad T = T_c) \tag{4.3.14}$$

$$P_2 = \frac{<O_i^4>}{zJ<\sigma_i O_i^3>} \quad (= 1, \quad T = T_c) \tag{4.3.15}$$

$$P_3 = \frac{<O_i^6>}{zJ<\sigma_i O_i^5>} \quad (= 1, \quad T = T_c) \tag{4.3.16}$$

$$P_4 = \frac{zJ<\sigma_i \tanh(\tfrac{1}{2}KO_i)>}{2<\sigma_i O_i>} \quad (= 1, \quad T = T_c). \tag{4.3.17}$$

Eqs. (4.3.14) – (4.3.16) are the $n = 0$, 1, and 2 versions of Eq. (4.3.9), and Eq. (4.3.17) is the criticality condition itself, Eq. (4.3.13). The zeroth-order equation involves only pair correlation functions and for later reference it is useful to write this out explicitly as

$$P_1(sc) = (\Gamma_0 + 4\Gamma_2 + \Gamma_4)/6\Gamma_1 \tag{4.3.18}$$

$$P_1(bcc) = (\Gamma_0 + 3\Gamma_2 + 3\Gamma_3 + \Gamma_5)/8\Gamma_1 \tag{4.3.19}$$

$$P_1(fcc) = (\Gamma_0 + 4\Gamma_1 + 2\Gamma_2 + 4\Gamma_3 + \Gamma_4)/12\Gamma_1 \tag{4.3.20}$$

$$P_1(diamond) = (\Gamma_0 + 3\Gamma_2)/4\Gamma_1. \tag{4.3.21}$$

Γ_0 is the auto correlation function, and $\Gamma_1, \ldots, \Gamma_5$ are the nearest-neighbor, $\ldots$, fifth-nearest neighbor correlation functions.

For the multi-spin correlation functions in Eqs. (4.3.15) – (4.3.17), there exists no finite-size scaling theory to predict the asymptotic N-dependence of the functions (cf. Sec. 3.1.3 for the pair correlation functions). Therefore, in analyzing the Monte Carlo results we have to resort to an empirical, although systematic, finite-size analysis. However, the convergence to the large-N limit is rather rapid and turns out to be facilitated by fortunate cancellation effects. The lattice sizes employed range from $N = 4 \times 4^3$ to 4×16^3 for the diamond lattice and from $N = 12^3$ to 20^3 for the simple cubic lattice. The statistics involves $2000N$ – $10000N$ Monte Carlo steps. The calculations are performed in a narrow temperature region around the critical temperature. The Monte Carlo data to be presented below are obtained using coarse-grained averages (cf. Sec. 2.2.7) of the P_i constructed from the coarse-grained averages of the numerators and denominators, separately. This calculation procedure allows for the statistical correlation between the fluctuations of the coarse-grained averages of the constituents, thereby reducing the statistical errors introduced by the use of finite ensembles.

In Fig. 4.3.1 are shown the results of P_1 and P_2 for the diamond lattice, and Fig. 4.3.2 gives the results of P_1, P_2, P_3, and P_4 for the simple cubic lattice. P_1 is found to have the most pronounced size dependence. In the critical region, P_1 decreases with increasing N. This tendency is in accordance with expectation since in Eqs. (4.3.18) – (4.3.21) the more distant correlation functions in the numerators are more overcorrelated (due to the periodic boundary conditions) than is the nearest-neighbor correlation function in the denominators. Nevertheless, the finite-size effects for the composite quantity P_1 are less dominant than those observed for the individual pair correlation functions, cf. Fig. 3.1.1. In going from P_1 to P_2 to P_3, the finite-size effects are reduced, most probably due to a progressive cancellation.

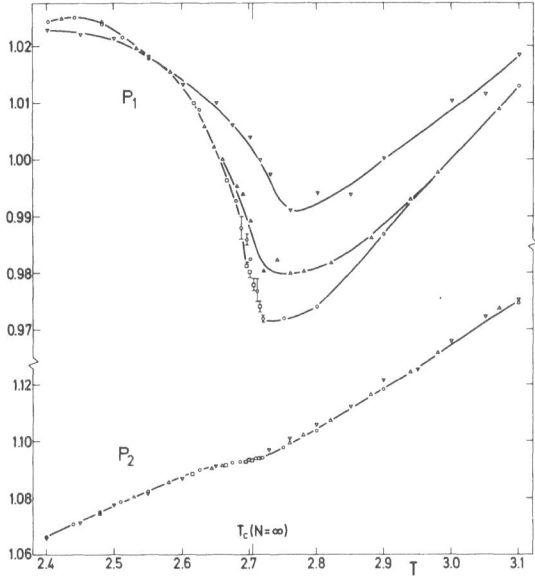

Fig. 4.3.1. Temperature dependence of P_1 and P_2, Eqs. (4.3.14) and (4.3.15), for the diamond lattice. The temperature is in units of J/k_B. $T_c(N = \infty)$ denotes the critical temperature of the infinite lattice (cf. Sec. 3.1.3). Monte Carlo data are given for lattices with $N = 4 \times L^3$ sites. $L = 4$ (inverse triangles), $L = 6$ (triangles), $L = 12$ (circles), and $L = 16$ (squares).

We now turn to a discussion of the Monte Carlo estimates of the critical values of P_1, P_2, P_3, and P_4 in relation to the basic assumption of the critical correlation theory of Sec. 4.3.1. The Monte Carlo results quoted below are derived at the critical temperature, T_c^S, given by series analysis. This is a consistent procedure since the critical temperature, T_c^F, determined from the critical correlation theory for the simple cubic lattice[*] only differs by 0.6% from T_c^S and since the P_i vary only slightly over that temperature range. It is immediately seen that none of the assumptions, $P_i(T = T_c) = 1$, is exact for either lattice. Nevertheless, it appears that the basic assumption may still be considered a reasonable approximation for the critical correlation functions. The approximation is better for the simple cubic lattice than for the diamond lattice. In the case of P_1, the Monte Carlo estimate is 0.978 ± 0.002 for the diamond lattice and 1.010 ± 0.003 for the simple cubic lattice. The latter result is in good agreement with the corresponding value 1.001 ± 0.007 obtained by insertion of the high-temperature series estimates (Ritchie and Fisher

[*] So far, the diamond lattice has not been treated within the critical correlation theory.

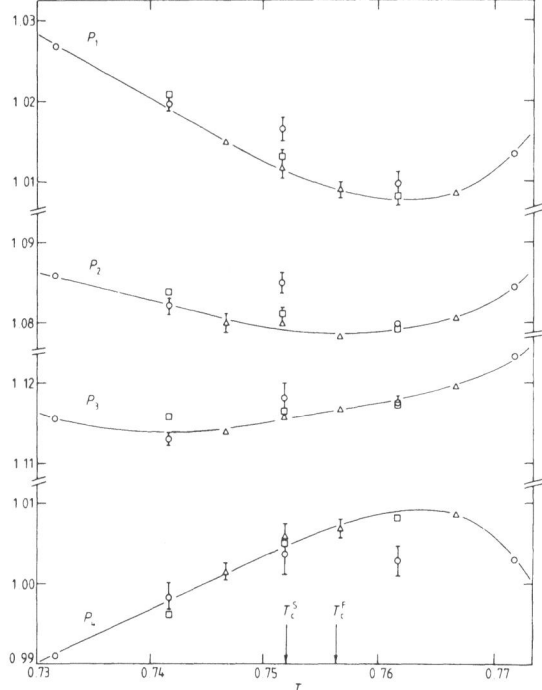

Fig. 4.3.2. Temperature dependence of P_1, P_2, P_3, and P_4 (cf. Eqs. (4.3.14) - (4.3.17)) for the simple cubic lattice. The temperature is in units of $3J/k_B$. T_c^S and T_c^F are the critical temperatures of the infinite lattice as estimated from series analysis (Sykes et al. 1972) and from critical correlation theory (Frank et al. 1982). Monte Carlo data are given for lattices with $N = L^3$ sites. $L = 12$ (circles), $L = 16$ (squares), and $L = 20$ (triangles).

1972) of the critical pair correlation functions into Eq. (4.3.18). Indeed, the zeroth-order approximation, $P_1 = 1$, seems to be extremely accurate for all three cubic lattices in that the series values of the critical correlation functions, via Eqs. (4.3.19) and (4.3.20), yield $P_1 = 1.006 \pm 0.010$, and $P_1 = 1.006 \pm 0.005$ for the body-centered and the face-centered cubic lattices, respectively. The basic assumption becomes less validated by the Monte Carlo data when going to P_2 and P_3 whose critical values are 8% and 12% off the assumption. The criticality condition, $P_4 = 1$, is validated to within $\frac{1}{2}\%$.

Although the basic assumption of the critical correlation theory of Sec. 4.3.1 is found not to be exact, the critical temperature calculations for the cubic lattices (Frank et al. 1982) demonstrate that the basic assumption is indeed very effective and leads to very accurate results. The Monte Carlo calculations presented in this section provide evidence that the critical correlation theory is founded on a sound physical basis and serve to indicate that the theory was on the right track. Subsequently, this has stimulated the development of a consistent theory (Frank and Cheung 1984) wherein the basic assumption has been modified. Therefore, the computer experiments reported in the present section may be considered a most successful type in the sense that they have stimulated a theoretical development and thereby made themselves superfluous.

5. Numerical Experiments

Out of the many different microscopic interaction models used in computer studies of phase transitions and ordering processes, only a small number of models are claimed to be quantitatively realistic models of specific physical systems. In this chapter, we shall present three of these models which are realistic in the sense that not only do they produce the correct symmetry properties of the various ordered phases but they also contain the appropriate energetics. Thus, the results of computer studies of these models often permit a quantitative comparison with laboratory experiments. In particular, Monte Carlo simulations of the models may be considered numerical experiments which provide additional experimental information.

The first model (Sec. 5.1) is a model of a wet bilayer aggregate of lipid molecules. Lipid bilayer systems, which are models of biological membranes, now appear to be sufficiently well-characterized experimentally to render it meaningful to attempt a quantitative description in terms of basic physical principles. The ultimate goal of such a program is to understand the relationship between the physiological *function* of a membrane and its *structure*. For that purpose, the lipid bilayer model is extended to include biologically important »impurity« molecules, such as cholesterol, polypeptides, and proteins. Since lipid bilayers and biological membranes are very seldom described in the physics literature, some space will be devoted here to introduce these systems. The reader, who prefers to proceed directly to the physical models of a membrane and the computer studies of these models, is suggested to move on to Sec. 5.1.6.

The second model (Sec. 5.2) is a model of nuclear spin systems (such as CaF_2, LiH, and LiF) coupled by dipolar interactions. The model gives rise to an extremely rich phase diagram including phases with nuclear dipolar magnetic ordering. This ordering usually takes place in the micro degree K region thus making experimental observations of the phase transitions and the ordered phases extremely difficult.

The third model is a model of the orientational properties of homopolar diatomic molecules (e.g. N_2, D_2, and H_2) adsorbed on the surface of graphite. The model may be readily extended to describe mixtures of diatomic molecules and rare gases (e.g. Kr and Ar). Two aspects of the model are studied in this chapter. In Sec. 5.3, the nature of the orientational ordering transition is investigated. It turns out that the model and its physical realizations are representatives of two universality classes (the Heisenberg model with cubic anisotropies) whose phase transitions have only been understood within the last few years. In Sec. 5.4, the model is used to study the domain-growth kinetics of systems with degenerate order parameters. This study illustrates the potentials in computer simulation of crystallization processes and the kinetics of growth.

5.1 Phase Transitions in Lipid Bilayers and Biological Membranes

5.1.1   What are biological membranes and what do they do?

All living cells are surrounded by a plasma membrane which is the most common cellular structure in animals as well as plants. A substantial fraction of biological activity takes place in association with the membranes, such as transport of matter, energy transduction, immune recognition, intercellular communication, nerve conduction, and biosynthesis.

Despite the fundamental importance of biomembranes to life processes, the work on a quantitative physical characterization of membrane structure has progressed rather slowly compared to work on other cellular components. However, in recent years some important progress has been made by supplementing the more traditional biochemical and biological approaches with modern and very powerful physical techniques whose potentials are well-known from condensed matter physics. Among these techniques are X-ray and neutron diffraction, solid state magnetic resonance (NMR and EPR), calorimetry, fluorescence anisotropy, circular dicroism, Raman scattering, and micromechanical measurements (for recent reviews, see e.g. Quinn and Chapman 1980, Israelachvili et al. 1980, Seelig and Seelig 1980).

The experimental efforts have provided us with the picture of a membrane as a pseudo-two-dimensional assembly of *lipid* and *protein* molecules forming, in an aqueous environment, a *bilayer* as shown in Fig. 5.1.1. The bilayer thickness ranges from 50 to 100Å. At physiological temperatures, the lipid matrix is a two-dimensional fluid in which the proteins float around. The lateral diffusion of the lipids is very fast ($D \simeq 10^{-8} \mathrm{cm}^2 \mathrm{sec}^{-1}$). This is contrasted by a very long (flip-flop) exchange time for moving lipid molecules from one side of the bilayer to the other ($\tau \simeq 10^7 \mathrm{sec}$). The low flip-flop rate serves to maintain the *bilayer asymmetry* which is of extreme importance for supporting the various functions of the membrane. Polar as well as non-polar lipids may occur, e.g. dipalmitoyl phosphatidylcholine (*DPPC*) and cholesterol. The lipids are amphiphilic molecules containing a water-

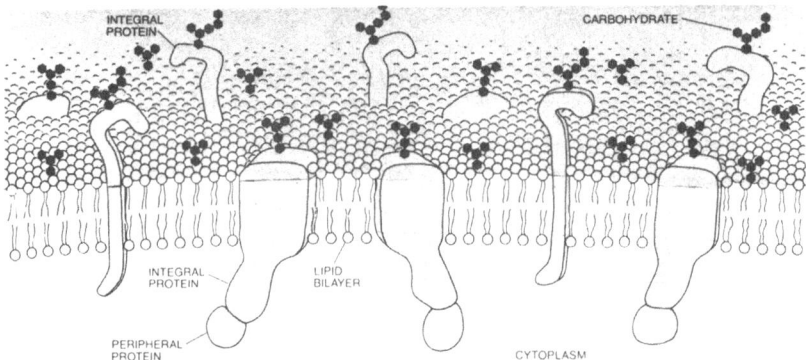

Fig. 5.1.1 Schematic model of a biological membrane. (Reproduced with permission from *Scientific American* (Lodish and Rothman 1979)).

soluble part (polar head) as well as a water-insoluble (hydrophobic) tail. The tail consists of one or two hydrocarbon chains. It is the amphiphilic property which is responsible for the spontaneous formation of aggregates (such as bilayers) of lipids when immersed in an aqueous medium (cf. detergents and soaps in water). The various polar lipids differ with respect to length and degree of saturation of the hydrocarbon chains as well as to type of polar head group. The lipid bilayer in itself presents a barrier impermeable to transport of hydrophilic molecules and ions across the membrane. It is the proteins associated with the membrane which are responsible for the various membrane functions mentioned above. There are basically two types of membrane-bound proteins: (i) peripheral proteins which are attached to the surface and (ii) intrinsic transmembrane proteins which span the entire bilayer thickness. In most natural membranes, the protein content is rather high, ~ 50wt%. Finally, a third membrane component, carbohydrates, should be mentioned. The carbohydrates are covalently attached to the proteins and protrude into the extracellular medium forming »social antennas« which may play an important role in intercellular recognition.

In contrast to what is the case for water-soluble proteins functioning in the aqueous medium, only limited information is available on the full three-dimensional structure of membrane-bound proteins. This is so because these proteins, in their capacity of being amphiphilic, are reluctant to form three-dimensional protein crystals which allow diffraction studies. So far, the structure has only been worked out for one protein, bacteriorhodopsin, which spontaneously forms two-dimensional crystals in its natural membrane (Henderson and Unwin 1975). Bacteriorhodopsin, which functions as a light-sensitive proton pump in the purple membrane of *halobacterium halobium*, is found to be an aggregate of seven hydrophobic α-helical polypeptide chains which traverse the membrane and are coupled together at the hydrophilic surface by hydrophilic amino acid residues. It is likely that the majority of transmembrane proteins have a similar structure and that the »elementary particles« of proteins are α-helical polypeptides.

5.1.2 Lipid bilayers are model membranes

A characteristic feature of natural biomembranes is the diversity of the components, e.g. the human red blood cell membrane is estimated to have several different proteins in addition to at least a couple of hundred different lipid species. Obviously, it is extremely difficult to interpret experimental measurements on such complicated systems lacking information on their exact composition, not to speak of the structure of many of the individual components. Naturally, this has led to studies of model membranes composed of a small number of lipid species and no proteins or with just a single type of protein (Seelig 1981). Such model membranes can be made either synthetically or by reconstituting native membranes. This development, in turn, has led to questions about the physical properties of the simplest possible biomembrane, a pure one-component lipid bilayer.

As we shall see, even this system is fairly complex from a physical point of view but may nevertheless offer the possibility of determining a number of relevant parameters in the description of real membranes.

5.1.3   Phase behavior of lipid bilayers

The type of system we will have in mind is a mixture of *DPPC* and water. The phase diagram includes several phases: the subphase (L_c), the gel phase ($L_{\beta'}$), the rippled phase ($P_{\beta'}$), and the fluid phase (L_α). In the case of excess water, which biologically is the interesting one, three successive first-order phase transitions are usually encountered as a function of temperature, namely the subtransition, the pretransition, and the main transition.

The structure of the various phases has been determined by X-ray diffraction (Ruocco and Shipley 1982, Ranck 1983). Only the upper (*main*) transition is thought to have biological significance since very small amounts of impurities (such as proteins) seem to eliminate the lower transitions. In what follows, we shall focus on the main transition.

The main transition is signalled by a sharp peak in the specific heat and an abrupt change ($\leq 20\%$) in the bilayer thickness leaving the total volume almost constant. The first-order nature of the transition is inferred from the presence of hysteresis. The entropy change is for *DPPC* about $\Delta S = 15 k_B$ per molecule which tells us that a very large number of degrees of freedom, $e^{\Delta S/k_B} \sim 3 \times 10^6$, is activated by the phase transition. By comparison, the liquid-gas transition of *Ne* has $\Delta S \sim 7 k_B$ per particle, the two-dimensional dislocation unbinding transition of Kosterlitz and Thouless has $\Delta S \leq 2 k_B$ per molecule (Doniach 1980), and smectic melting has $\Delta S \sim \frac{1}{2} k_B$ per molecule. Thus, we are faced with the striking result that the lipid phase transition is orders of magnitude more entropic than classical three-dimensional liquid-gas transitions or fashionable two-dimensional melting transitions. Obviously, the source of the entropy in the lipid system is the rotational isomerism of the hydrocarbon chains. It is now commonly believed that the main transition is a consequence of *chain melting* where the individual lipid chain goes from an extended low-temperature conformational state to a highly disordered high-temperature state, subject to the constraint of the polar heads being confined to the bilayer surface. The transition bears some resemblance to polymer melting.

5.1.4   Back to biology: Are phase transitions at all relevant to the biological functions of the membrane?

Most living organisms function at a fairly constant temperature. Membrane phase transitions and changes in the thermodynamic state of the lipids would therefore be of little importance if they could be triggered by temperature changes only. However, phase changes can also be induced by alterations in the chemical composition of the membrane as well as by appropriate environmental variations, such as changes in pH, ionic strength, and the electric field across the membrane. It is important to stress that when speaking about »phase transitions« in membranes, one should rather say »phase behavior« and consider the diversity of the phase diagram of a many-component system. Thus, phase coexistence will often occur and sharp phase changes are restricted to very special cases. Some real cell membranes do, however, exhibit rather clear phase transitions. For example, the native membranes of the primitive bacterium *Acholeplasma Laidlawii* has a phase change in a narrow temperature range, as indicated by an abrupt change in the moments of the ^2H-magnetic resonans spectra (Davis et al. 1980).

So far, no clear-cut demonstration has been given of a direct influence of biomembrane phase transitions on the physiological functions of the membrane. Nevertheless, a number of strong indications have been reported: Most living cells require a fluid membrane in order to function, and it is generally observed that the growth temperature is above the phase transition temperature. Bacterial membranes seem to be in the transition region whereas mammalian membranes are safely fluid. As an interesting example it may be mentioned that certain warmblooded animals which hibernate during winter maintain their membrane fluidity by altering the lipid composition. The activity of many membrane-bound enzymes and proteins is abruptly altered close to lipid phase transitions. The enzyme function may thus be controlled by appropriate phase changes. In particular, the fluidity of the lipids surrounding the enzyme and the matching of the hydrophobic regions appear to be important parameters.

A number of speculations can be presented which indicate that it is favorable for a membrane to be close to a phase transition: The energy and the lateral density fluctuations are strongly enhanced near the lipid transition (the lipid bilayer may indeed be close to a critical point!). This leads to a corresponding enhancement of the passive transmembrane permeability of matter, as invariably observed by experiments. The large lateral density fluctuations imply a large compressibility and an increased probability for pore formation. Phase separation leads to the possibility of forming specific (local) environments for proteins, which will facilitate their functioning. Specifically, protein aggregation can occur which may be important for endocytosis, formation of transmembrane channels, and for intercellular communication. The action of anaesthetics and antibiotics in very low concentrations is enhanced by phase separation. The phase separated membrane contains domain boundaries, at which the lipid packing is less effective, and thus has a large capacity of incorporating new components. The extent to which lipids and proteins interact is still a matter subject to much dispute. Most proteins require a fluid lipid environment to function, but no lipid specificity has been found. The outer regions of the proteins appear to be fluid themselves and the direct lipid-protein interaction may very well be an intimate matching of the low-frequency dynamics. Conformational changes in the proteins and thus in their function can be induced by changes in hydrophobic thickness of the surrounding lipid matrix.

5.1.5 Theories of lipid bilayer phase transitions

Seen from a theoretical physics point of view the phase transition in lipid bilayers is not very exotic or exciting compared to two-dimensional melting transitions, critical phenomena in magnetic systems, superfluid transitions, or transitions in gauge theories. First of all, this is due to the fact that we are dealing with a first-order transition. Secondly, the transition may not even be driven towards a critical or tricritical point since it is not physically possible to apply an external lateral pressure to the bilayer. (In constrast, this can be done in the case of lipid *monolayers* on e.g. a water surface (Albrecht et al. 1978).) Thus modern concepts of universality cannot come into play. Still, the lipid transition is a very striking physical phenomenon, and it is a challenge to theory (i) to identify the source of entropy involved in the transition, (ii) to determine the relevant interactions of the system, (iii) and to calculate phase diagrams for lipid systems in the presence of biological important impurities such as proteins (for recent reviews on theories of the main lipid bilayer phase transition, see Nagle 1980, Caillé et al. 1980, Bell et al. 1981, Pink 1983).

Basically, the theory has progressed along two different routes using respectively (i) phenomenological (Landau-type) models and (ii) microscopic interaction Hamiltonians with different degrees of realism. Except for extremely simplified cases (Nagle 1975), approximate solution schemes have to be introduced in order to calculate the physical properties of the microscopic models. In most cases, the methods of solving the models barely go beyond the mean-field approach. However, recently several computer studies of very realistic microscopic models have been carried out, and we shall report on these in the following sections.

When constructing a theoretical model of a lipid bilayer, the following points should be taken into consideration: (i) the lipid chains are subject to rotational isomerism, (ii) there is no covalent character in the interaction between the lipid molecules, (iii) the chains interact via anisotropic van der Waals dispersion forces, (iv) the polar heads interact via central polar forces (coulomb, dipolar), (v) excluded volume interactions are effective, and (vi) water and interaction between bilayers may play a role.

A remark is in order on the effective dimensionality of lipid bilayer membranes (Doniach 1980, Nagle 1980). This point is of interest since it might be thought that the lipid bilayer transition is an excellent candidate for two-dimensional melting in the Halperin-Nelson sense (Halperin and Nelson 1978). This is not so, however, basically because the lipid bilayer transition proceeds via the internal degrees of freedom of the hydrocarbon chains and not via translational coordinates. Spatial molecular disorder is expected to play only a minor role in lipid bilayer melting and it contributes very little to the large entropy of melting.

5.1.6 Computer simulations of lipid bilayers

It is obvious from the above introduction to membranes, that it is a very hard problem to make good theories of lipid bilayers, not to speak of natural biological membranes. It is feasible to write down a very general interaction model Hamiltonian of a lipid bilayer taking into account the points (i) - (vi) listed in the preceeding section. However, this general model would be so complicated that there would be very little hope of calculating any of its properties. One possibility, then, is to simplify the model to a tractable form at the expense of the degree of realism of the model. In the extreme case, the simplified model may turn out to be exactly solvable (Nagle 1980). Such extremely simplified models may serve as guides for determining the relevant ingredients of more realistic models. If, however, the objective of the theoretical approach is to describe quantitatively and in more detail the wealth of experimental information which is becoming available these years, and if the theorist shall interact fruitfully with the experimentalist by proposing interesting and relevant experiments, there is a need of models with a high degree of realism.

It is at this point, modern computer simulation techniques become extremely valuable. These techniques may readily be applied to very complicated models without introducing further approximations. The many advantages of such techniques should sufficiently outbalance the drawback that »they involve a large amount of computer processing, and this tends to obscure their physical meaning« (de Gennes).

A number of numerical simulations have in recent years been carried out on microscopic interaction models of lipid bilayers using molecular dynamics as well as Monte Carlo methods. The most realistic molecular dynamics calculation on lipid bilayers carried out so far is probably that of van der Ploeg and Berendsen (1982) (see also Frischleder 1981). Using small systems of 2 × 16 and 2 × 64 decanoate molecules subject to Lennard-Jones, dihedral, and bond angle potentials, these authors studied the fluid region and were able to reproduce the experimental order parameter profile. Moreover, they found that the characteristic plateau in the profile (Seelig and Seelig 1980) is due to a slowly fluctuating, collective tilting mode in the chains. These findings are in accordance with the molecular dynamics calculations performed by Kox et al. (1980), who on the one hand used a somewhat less realistic potential for a monolayer, but on the other hand were able to observe indications of a first-order phase transition from an ordered fluid-like state to a disordered gas-like state. The two-dimensional dumbbell model advanced by Cotterill (1976), which from molecular dynamics simulations is shown to exhibit a very interesting disclination-mediated melting transition, is hardly a proper model of the main lipid bilayer transition since it completely neglects the rotational isomerism of the chains. However, this model may be pertinent for the recently discovered subtransition (Chen et al. 1980).

The earliest Monte Carlo studies of lipid systems were simulations of lipid hydrocarbon chain conformer formation (Wittington and Chapman 1966, Scott 1977, Scott and Cherng 1978). These studies were only able to describe very small chain clusters ($\sim$ 10 chains) because they treated the chain conformers in some detail. A hard core barrier representing a cholesterol molecule were also included in these simulations. The order parameter »plateau« was obtained as a function of lateral packing density, but these models could not describe cooperative effects involving many chains, such as the lipid chain melting transition. The Monte Carlo calculations on the multi-state models of lipid bilayers to be reported below are virtually the first simulations which have provided some detailed insight into the main lipid bilayer phase transition.

5.1.7 Multi-state models of lipid bilayers

The q-state model of lipid bilayers is the condensed-matter physicist's model of a complicated physico-chemical system. The phospholipid bilayer is considered as two lipid monolayers which in general are assumed to be independent of each other. The model is designed to describe only the main gel-to-fluid transition and not properties of the rippled $P_{\beta'}$ phase. Since translational degrees of freedom are irrelevant for the main transition, each monolayer is represented by a two-dimensional triangular lattice with N lattice sites. The lattice approximation automatically takes care of the excluded volume interactions. Each site of the lattice is occupied by a single saturated hydrocarbon chain of the phospholipid molecule and is composed of M monomers. Each chain can exist in one of q states and each state m is described by a cross-sectional area A_m, an internal conformational energy ϵ_m, and a degeneracy D_m. D_m forms the single-chain density of states for the model. All q states may be derived from the ground state, the all-*trans* state ($m = 1$), by rotational isomerism. The areas of the states are constructed so that the volume per chain is conserved (Caillé et al. 1980, Marcelja 1974). The energy spectrum is bounded upwards due to the presence of a highly excited »fluid« state ($m = q$) modelling the »melted« state of a chain.

The Hamiltonian for the q-state model consists of three parts:

$$H = H_o + H_c + H_p, \tag{5.1.1}$$

where the single-site energy is given by

$$H_o = \sum_{j=1}^{N} \sum_{m=1}^{q} \epsilon_m L_{mj}. \tag{5.1.2}$$

L_{mj} is a projection operator for a hydrocarbon chain in the mth state on site j. H_c describes the anisotropic van der Waals interaction between the hydrocarbon chains in terms of their areas

$$H_c = -\frac{J_o}{2} \sum_{<i,j>} \sum_{m,n=1}^{q} I_m^c I_n^c L_{mi} L_{nj}, \tag{5.1.3}$$

where

$$I_m^c = w_m (A_1/A_m)^{5/4} \sum_p S_{mp} / \sum_p S_{1p}. \tag{5.1.4}$$

w_m is a set of model parameters, and S_{mp} is the nematic order parameter for the pth $C - C$ bond of the hydrocarbon chain (Seelig and Seelig 1980). The interaction is spatially isotropic and the strength is given by J_o. Only interactions with the six nearest neighbors in the triangular lattice are taken into account. Finally, H_p accounts for the interaction between the polar heads of the phospholipid molecules. As first suggested by Marcelja (Marcelja 1974), this interaction may be expressed in terms of an effective surface pressure Π

$$H_p^{(1)} = \Pi \sum_{j=1}^{N} \sum_{m=1}^{q} A_m L_{mj}. \tag{5.1.5}$$

The interaction between the polar heads can be more explicitly accounted for by the Hamiltonian

$$H_p^{(2)} = -\frac{K_o}{2} \sum_{<i,j>} \sum_{m,n=1}^{q} I_m^p(\nu) I_n^p(\nu) L_{mi} L_{nj}, \tag{5.1.6}$$

where

$$I_m^p(\nu) = (A_1/A_m)^{\nu/4} \tag{5.1.7}$$

and ν is the exponent of the central force between the polar heads.

The discreteness of the q-state model as formulated above is justified by the rotational isomeric model, known from polymer science, where the continuum of rotational angles is replaced by one *trans* and two *gauche* angles corresponding to the three minima of the rotational energy function. The two degenerate gauche states have conformational energies about $\epsilon = 0.45 \times 10^{-13}$erg above the *trans* minimum. This reduces the number of possible single chain states to $\sum_m D_m \sim 3^{M-1}$. In a condensed interacting system of chains, only a small number of these states will be accessed. The basic idea is therefore to choose q as a small number by selecting the most important states. Two states which always occur are the all-*trans* ground state and a high-energy »melted« state which is characteristic for the disordered state. Excited intermediate gel-like states are chosen by requiring low conformational energy and optimal packing. The values of A_m and D_m are determined by geometrical and combinatorial considerations, respectively (Caillé et al. 1980). This discrete model may be generalized to a continuous model which has not been studied so far. For the construction of the general model, the complete single-chain density of states determined by Day and Willis (1981) will be useful.

The discrete q-state model is a generalization of the q-state Potts model and was first proposed as a lipid bilayer model by Doniach ·(1978) and Caillé et al. (1978) for $q = 2$. The crucial difference between the $q = 2$ state model and the standard spin - $\frac{1}{2}$ Ising model is that the former model is associated with single-site degeneracies. It can be shown that the $q = 2$ state model can be mapped on a spin - $\frac{1}{2}$ Ising model in a temperature-dependent magnetic field (Pink et al. 1980a). Thus, the model is seen (from the exact Onsager-Yang solution) to describe, at least qualitatively, the first-order gel-fluid phase transition provided that the melting temperature is below the critical point. Pink and coworkers (Pink et al. 1980b, Caillé et al. 1980) have extended the model to include $q = 10$ states by introducing eight intermediate states which turn out to be important for describing optical Raman band intensities.

The thermodynamic properties of the $q = 10$ state model have been derived within the framework of the mean-field theory and the Oguchi pair approximation (Caillé et al. 1980). The model, in conjunction with these methods of solution, has proved very successful in reproducing a great variety of experimental measurements on pure lipid bilayers, such as Raman scattering intensities, ^2H-NMR order parameters, passive ionic transbilayer permeability, and transition temperatures and enthalpies (Caillé et al. 1980).

In view of the success obtained by the $q = 2$ and $q = 10$ state models in describing a variety of properties of biological membranes, it is of interest to study for these models the influence of fluctuations which are completely neglected in the mean-field approximation. For a large class of well-known spin models, it is known that the mean-field theory in the most favorable cases leads only to a qualitatively correct description of thermodynamic quantities, e.g. transition temperatures are typically off by a factor of two, and properties close to phase transitions, especially fluctuation quantities, are highly incorrect, not to speak of the finer details inherent in critical behavior, such as critical exponents. Now, why is it that both models within the mean-field picture lead to such an accurate description of lipid bilayers? Is it because the models contain so many parameters that, even with a very incomplete method of solution, they can be parametrized to experimental data for several different quantities? Or, is the mean-field solution for these particular models not so bad an approximation after all?

It is obvious that Monte Carlo simulations of q-state models are useful to shed light on these questions. In fact, the Monte Carlo technique seems well suited to do so as, contrary to other theoretical approaches, it may easily treat complicated Hamiltonians such as those in Eqs. (5.1.1) – (5.1.7). Furthermore, Monte Carlo calculations, having available all microscopic variables, provide detailed information on the molecular level, thus allowing several physical properties to be determined which are not easily accessible in any ordinary experiment.

5.1.8   Computer simulations of the q-state models for the gel-fluid phase transition

A detailed Monte Carlo study of the $q = 2$ and $q = 10$ state models have been reported by Mouritsen et al. (1983b). Due to the complicated form of the single-chain density of states, the standard realization of the Monte Carlo importance sampling procedure cannot be directly applied to the q-state models. Consequently, the generalization given in Sec. 2.2.3 is used. The transition matrix $\overline{\overline{P}}^{*}$ of Eq. (2.2.9) is chosen according to a sequential visitation of lattice sites, and the trial state of the site under consideration is chosen randomly among the q possible states. The Monte Carlo calculations are performed on rectangular-shaped triangular lattices with toroidal periodic boundary conditions and $N = L \times L$ lattice sites, $L = 30, 40, 60$, and 90. Each site is associated with a single hydrocarbon chain assigned a label m which describes the microscopic state of the chain. The basic microscopic variable is then the area, A_m. This area does not play the role of a real area in the model simulations; there is no free volume and the lattice is perfectly rigid. However, the thermal average of the area per chain determines the actual temperature dependent surface area of the bilayer, subject to the constant volume approximation. Thus, the temperature dependence of the thermal average of the area may be thought of as characterizing a temperature dependence of the lattice constant of the underlying triangular lattice.

The temperature dependence of several static thermodynamic properties is calculated, such as the chain cross-sectional area, the internal and free energies, the coherence length, the lateral compressibility, and the specific heat. Furthermore, the occupation variable of each state is determined. From the occupation variables, the ^2H-NMR order parameters and the Raman band intensities may be derived. Model parameters pertinent to dipalmitoyl phosphatidylcholine ($M = 16$) are chosen (Mouritsen et al. 1983b). We find the same results for the model with the pressure term, $H_p^{(1)}$ Eq. (5.1.5), and for the model with the direct interaction

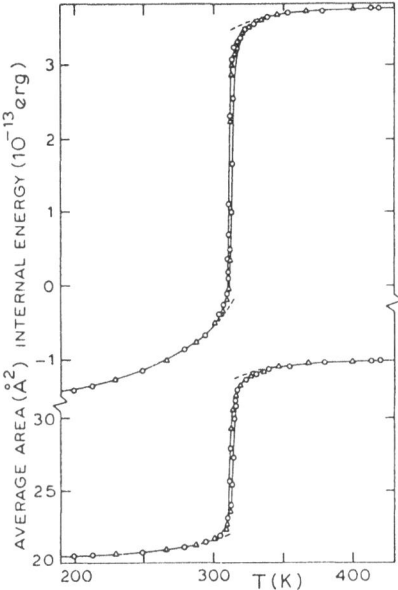

Fig. 5.1.2. Temperature dependence of the internal energy per chain and the average cross-sectional area for the $q = 10$ state model of a lipid bilayer. The data points are obtained from Monte Carlo calculations on systems with N chains. Circles: $N = 900$ and triangles: $N = 1600$. The dashed lines give the mean-field predictions. The transition temperature of the model is $T_m = 314K$.

between the polar heads, $H_p^{(2)}$ Eq. (5.1.6) with $\nu = 1$ (Coulomb forces). Naturally, the identity of the results encompasses only derivatives of the free energy F and not F itself. This demonstrates that, as for describing the main transition in the lipid chain reservoir, the polar head interactions are effectively and accurately accounted for by the surface pressure term. As an example, Fig. 5.1.2 shows the temperature dependence of the internal energy and the average cross-sectional area per chain for the $q = 10$ state model. Both properties exhibit hysteresis in a narrow temperature region around $T = 314K$, demonstrating that the phase transition of the model is of first order. The width of the hysteresis is around $1 - 2$ degrees. However, the transition is gradual (continuous) for both heating and cooling series of temperatures and takes place over approximately $4 - 5$ degrees. The gap between the high- and low-temperature branches for both $E(T)$ and $A(T)$ increases with N; however, the results for the two largest systems, $N = 3600$ and $N = 8100$, are the same within the statistical errors and therefore constitute a good approximation to the infinite lattice values. This behavior should be contrasted to that of the $q = 2$ state model which has a sharp first-order transition with jump-like behavior at the metastable branches (Mouritsen et al. 1983b).

Further evidence of the first-order nature of the transition for $q = 10$ may be obtained from the free energy function. The free energy is difficult to derive from Monte Carlo calculations because it basically requires an evaluation of the partition function itself. Here, we determine the free energy function for the two phases

from a numerical integration of the internal energy, choosing appropriate high-
and low-temperature boundaries for the integration procedure. From the first law
of thermodynamics and the definition of the free energy F, we immediately arrive
at the following expressions (cf. Eq. (2.2.28))

$$F(T) = \frac{T}{T_1} F(T_1) - T \int_{T^{-1}}^{T_1^{-1}} E(x)\, dx, \quad T_1 < T_m \tag{5.1.8}$$

for the low-temperature (gel) phase, and

$$F(T) = \frac{T}{T_2} F(T_2) + T \int_{T_2^{-1}}^{T^{-1}} E(x)\, dx, \quad T_2 > T_m \tag{5.1.9}$$

for the high-temperature (liquid crystalline) phase. Since mean-field theory is exact
for our model in the limits $T \to 0$ and $T \to \infty$ and since it moreover constitutes a
very good approximation for temperatures far from the transition region, we need
not work out the high- and low-temperature perturbation expansions in order to
determine the limits $F(T_1)$ and $F(T_2)$ but may obtain these directly from the mean-
field theory. Choosing the values $T_1 = 200K$ and $T_2 = 400K$, we have obtained
the free energy curves in Fig. 5.1.3 from Eqs. (5.1.8) and (5.1.9) by using the
internal energy data of Fig. 5.1.2. Only Monte Carlo data representing the large-
N limit have been employed in the integration. The discontinuous change in the
slope of $F(T)$ at the transition point demonstrates the first-order nature of the
transition. From this discontinuity, the heat of melting can be determined.

   We shall now in detail discuss the very nature of the lipid bilayer phase transition
on the basis of computer simulations of q-state models (Mouritsen 1983a). This
serves as an instructive illustation of how numerical »experimental« information,

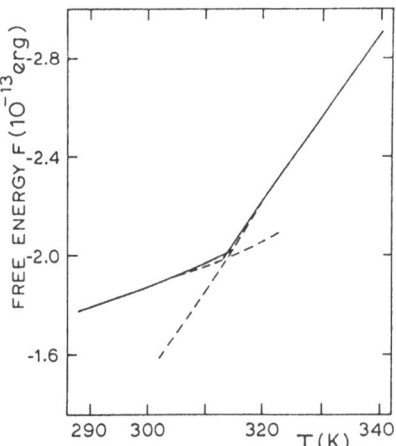

Fig. 5.1.3. Free energy vs temperature for the $q = 10$ state model of a lipid bilayer. Monte
Carlo (solid line) and mean-field (dashed line) calculations. The transition temperature is
$T_m = 314K$ in both descriptions.

interpreted in terms of events on the molecular level, can be useful to resolve puzzles posed by real experiments.

Since the pioneering work performed by Chapman and coworkers (Chapman et al. 1967), it has remained an open question why the endothermic gel-to-fluid first-order phase transition of pure one-component uncharged lipid bilayers appears as a broadened »continuous« melting process in every experimental investigation. It seems experimentally well established that this main transition is indeed of first order. In some experiments, the first-order nature is inferred from the presence of hysteresis. In other experiments, a complete reversible behavior is found and the first-order nature is assessed by a simple van't Hoff analysis. Most frequently, the »continuous« nature of the transition has been explained in terms of »lack of perfect cooperativity,« the degree of cooperativity being diminished by the presence of site defects or by small amounts of impurities. Obviously, it is of interest to determine the origin of the rounding of the first-order transition. It should be stressed at this point that for a pure one-component system in thermal equilibrium, Gibbs's phase rule forbids rounding of a transition and coexistence of phases over a finite temperature range. Such phenomena are therefore strictly non-equilibrium effects. The question we seek to answer here is whether there exists an intrinsic molecular property of the pure lipid bilayer which may be responsible for the observed »continuous« melting. If this is in fact the case, one need not look for more complicated explanations (e.g. in terms of impurities or experimental inadequacies) of the broadened transition.

The results reported above for the $q = 2$ and $q = 10$ state models suggest that a systematic study of the *appearance* of the phase transition in a computer simulation study as a function of q may provide some important insight into the nature of the transition. In such a computer study there is complete control over parameters (such as the value of q) which cannot be controlled in any real experiment. Furthermore, a system is studied which per construction is a genuinely pure one-component lipid bilayer with no lattice defects or impurities. The q-state models are here (Mouritsen 1983a) derived from the $q = 10$ state model by deleting an appropriate number of intermediate states, keeping constant the total number of states (i.e. the sum of the degeneracies).

In Fig. 5.1.4 are given the results for the average cross-sectional area per chain, $A(T)$, as a function of reduced temperature for various values of q. Results from increasing as well as decreasing temperature series are shown. The temperature has been changed in very small steps in order to facilitate the relaxation towards thermal equilibrium. In these calculations, a system with 3600 chains (with periodic boundary conditions) has been used. Calculations on a smaller system with 1600 chains yield the same results for the equilibrium values of $A(T)$. Figure 5.1.4 shows that for low values of q ($q \lesssim 5$) a clear hysteresis behavior occurs. The width of the hysterises as well as the discontinuities at the end points of the metastable branches decrease when more intermediate states are specified. For $q \gtrsim 6$, the hysteresis has disappeared and $A(T)$ varies continuously through the transition. For each temperature value studied, the system has been equilibrated for a very long time corresponding to about 10^4 excitations per chain. If the transition is scanned using shorter equilibration times (or larger temperature steps), *closed*

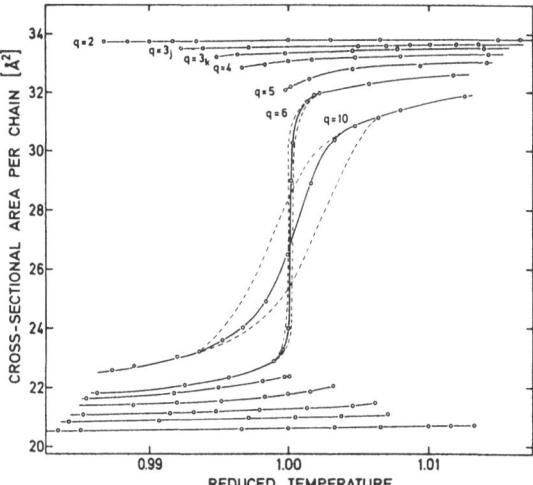

Fig. 5.1.4. Average cross-sectional area per chain as a function of reduced temperature, T/T_m, for q-state models. Circles indicate results derived from Monte Carlo calculations. For $q = 3$, the subscript refers to kink (k) and jog (j) chain states. The dashed loops specify for $q = 6$ and $q = 10$ the range over which metastable states have been detected.

hysteresis loops are observed for $q \gtrsim 6$, as indicated by the dashed loops in Fig. 5.1.4. Obviously, the system is relaxing extremely slowly in the transition region. The slow relaxation is also indicated by the fluctuation quantities, such as lateral compressibility and specific heat, which are found to have pronounced peaks at T_m. The peak intensities increase with the value of q, implying a softening of the system and an enhancement of the density fluctuations.

By an analysis of the microscopic chain configurations, we may gain insight into the microscopic phenomena which accompany the strong spatial density fluctuations in the transition region. An example of such configurations is given in Fig. 5.1.5 for the $q = 6$ state model. The figure demonstrates very clearly the formation of large *clusters* of correlated chains (fluid or gel-like) and the increase of cluster size when approaching T_m from both sides. For $T \lesssim T_m - 0.3$ °C, the system is predominantly gel-like and the metastable fluid phase is present as a number of isolated clusters consisting of $\lesssim 100$ chains. Conversely, for $T \gtrsim T_m + 0.3$ °C the bilayer is fluid with gel-like clusters floating around as isolated domains with $\lesssim 100$ chains. For $|T - T_m| < 0.3$ °C, huge clusters are formed encompassing $\gtrsim 500$ chains; the phase of the system is not well-defined and in the course of the time-development of the chain configurations the fluid and gel phases alternate in dominance, as exemplified in Fig. 5.1.5. Thus, the two phases effectively coexist over a *finite* range of temperatures. This is not the true thermodynamic equilibrium situation for the system and the result therefore emphasizes the extremely slow relaxation towards equilibrium. Increase of q leads to formation of larger clusters subjected to stronger fluctuations in size, i.e. the system softens. The relaxation towards equilibrium is slowed down by the intermediate chain conformational states which are found to have a tendency to appear

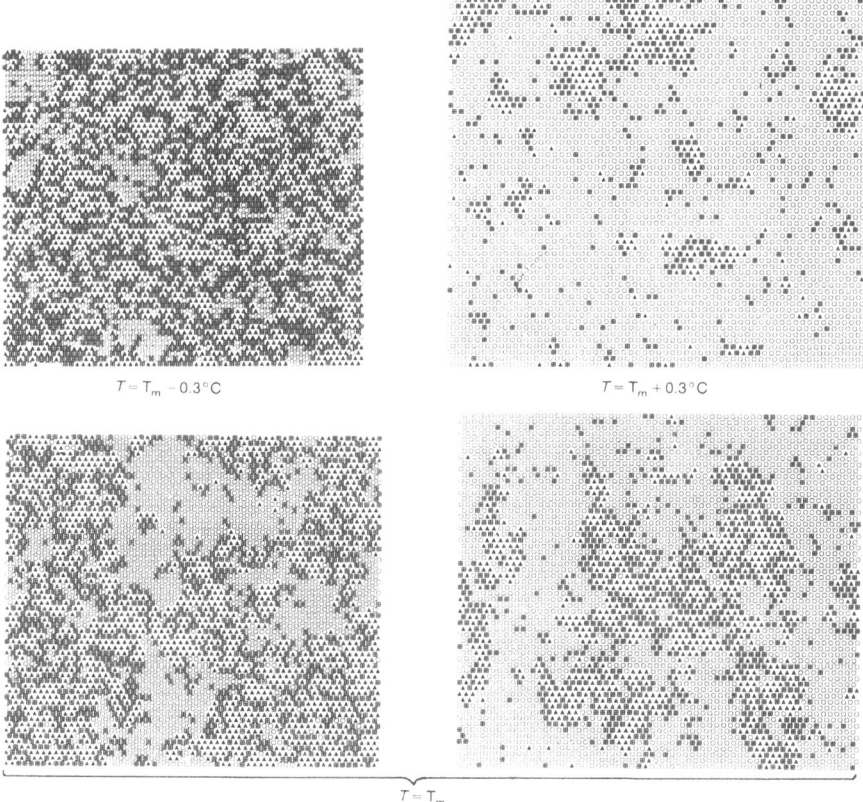

$T = T_m - 0.3\,^\circ C$

$T = T_m + 0.3\,^\circ C$

$T = T_m$

Fig. 5.1.5. Snapshots of characteristic microscopic configurations for the $q = 6$ state model. Configurations are given for temperatures immediately below, at, and immediately above the equilibrium melting temperature, T_m. The system consists of 3600 lipid chains on a triangular lattice. The lattice parameter has been scaled with the square-root of the average cross-sectional chain area to display the lateral expansion of the bilayer during the melting process. The symbols denote the conformational states of the lipid chains: solid triangles denote the all-*trans* state; squares with a cross denote intermediate states; circles denote the fluid (melted) state.

on the surface of the clusters, forming a very »*soft*« *domain wall* (»Bloch wall«) which coats the clusters. The softening of these walls decreases the ratio between cluster surface energy and bulk energy. This in turn leads to a screening of the interaction between the clusters and effectively slows down the tendency of the clusters to fuse and to form a bulk phase.

Recasting these observations in phrases used for describing the kinetics of phase transitions (Penrose 1978), three stages of the melting process can be described: (i) the *nucleation* occurs very rapidly mediated by intermediate chain states which decrease the surface energy of the nucleation centers; (ii) the *growth* of the nucleation centers is also fast and is driven by the internal entropy increase resulting from »melting« of the individual chains into the fluid state; and (iii) the *coarsening* (or the fusion of clusters) is dramatically slowed down by the low surface energies.

The last step therefore becomes the rate-determining one. In other words, a large number of intermediate chain configurations facilitate a kinetic stabilization of metastable cluster distributions. The observable macroscopic consequence of this phenomenon is a »continuous« melting transition.

This finding of a smeared first-order transition in uncharged lipid bilayers is in excellent agreement with experiments on a variety of lecithin bilayers. Furthermore, recent freeze-fracture work[*] on lipid vesicles has revealed a domain structure in the transition region similar to that shown in Fig. 5.1.5.

The passive transbilayer permeability of the lipid bilayer to ions (e.g. K^+ and Na^+) and small molecules (e.g. anaesthetics) is directly related to the density fluctuations (Doniach 1978). The present calculations therefore predict a strong enhancement of the permeability in the transition region in accordance with experiments. Moreover, in terms of the cluster picture it can be argued that matter is likely to permeate through the bilayer in connection with the soft domain walls where the packing of the lipid molecules is less effective. By reasoning along the same lines, it may be suggested that other components of lipid bilayer membranes, such as cholesterol and small proteins and polypeptides, are likely to be positioned in the domain walls which will offer the softest environment and require the smallest expense in packing energy. These molecules will pin the domain walls and slow down the relaxation to equilibrium even further. If we neglect phase separation, this will then lead to a further enhancement of the static part of the density fluctuations and thus of the compressibility and the passive permeability, as invariably found by experiments. The potential biological importance of such phenomena has been discussed by Lee (1977) and Sandermann (1978).

The q-state models studied here are also believed to be realistic models of lipid monolayers (Caillé et al. 1980, Georgallas and Pink 1982). Therefore, the results described here may equally well offer an explanation of the non-horizontal isotherms found for the liquid-condensed to liquid-expanded transition in lipid monolayers (Albrect et al. 1978, Cadenhead et al. 1980).

The computer studies on the lack of cooperativity in the lipid bilayer melting process suggest that very interesting kinetics are associated with this transition. Experimentally, there is currently a lot of interest and activity associated with the phase transition kinetics of lipid bilayers as revealed by pressure- and temperature-jump techniques (Elamrani and Blume 1983, Tsong and Kanehisa 1977). The use of X-ray synchrotron radiation is expected to open up new possibilities for quantitative studies of the dynamics associated with the bilayer melting process (Ranck 1983). In these experiments, the relaxation towards equilibrium following a sudden change in either hydrostatic pressure or temperature is studied as a function of time. Several different relaxation times can be resolved. To provide guidance in the interpretation of these specific relaxation phenomena, we have undertaken a kinetic study of the $q = 6$ state model. The model is quenched from just above to just below (and *vice versa*) its transition region ($\Delta T \simeq 0.6$ °C) and the relaxation towards equilibrium is recorded as a function of Markov time. A typical plot of the evolution of the coarse-grained averages of the cross-sectional

---

[*] Eric Sackmann, private communication.

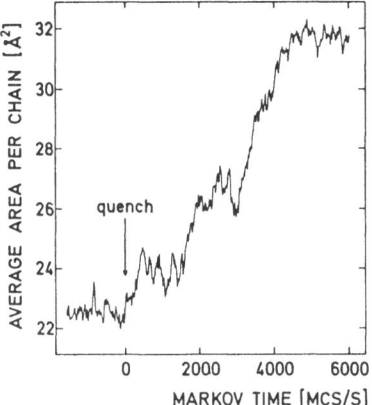

Fig. 5.1.6. Time evolution of the coarse-grained averages of the cross-sectional area per chain for the $q = 6$ state model following a quench in temperature from immediately below to immediately above the transition region. The Markov time is in units of Monte Carlo steps per site.

area per chain is shown in Fig. 5.1.6. The data in this figure are obtained from a system with $N = 3600$ chains. The relaxation proceeds as a cascade via several intermediate metastable states corresponding to the plateaus in Fig. 5.1.6. The cascading relaxation gives rise to a family of well-resolved peaks in the distribution function for $A(T)$. Each peak corresponds to a horizontal portion in Fig. 5.1.6. The relaxation appears to be asymmetric and dependent on the direction of the quench. From snapshots such as those shown in Fig. 5.1.5, it is found that the various metastable states are characterized by different cluster distributions, the average size of the fluid clusters increasing jumpwise when the system escalates the staircase after an upwards temperature quench. These metastable states are presumably the same as those encountered in attempts to equilibrate the q-state models in the transition region. It is noteworthy that for $q < 5$ cascading is not found. Thus, the observations of cascade kinetics and a smeared first-order transition are interrelated and consequences of the same molecular property, namely the large number of intermediate chain configurations. A completely analogous staircase shape of the relaxation is found for the coarse-grained internal energies. The exact shape of the cascade and the number of resolved intermediate steps are not reproducible, but depends on the details of the quench, such as initial configuration, random number sequence, and the change in temperature, However, a cascading relaxation is always found and is therefore established as a real property of the model. Consequently, the q-state model is also found to be, at least qualitatively, a proper model of the kinetics of lipid bilayer melting. Since it is not feasible to relate directly the Markov time to real time, we are unable to provide a quantitative comparison of the relaxation times in the simulations with those found in the real experiments.

We conclude the section on the computer simulations of the q-state models of pure lipid bilayers by the following remarks on the model parameters: For both the $q = 2$ and $q = 10$ state models it is found, by comparison with the corresponding mean-field properties, that all thermodynamic properties, and especially the response functions, are strongly influenced by thermal density fluctuations in a region of about ±5 degrees around the phase transition at $T_m = 314K$. This has the significant implication, that if mean-field solutions for either model are used to fit the experimental values of the transition temperature and the heat of melting, the numerical values of the interaction parameters required for the analysis would be incorrect. Furthermore, predictions based on q-state models of the behavior of most thermodynamic quantities of lipid bilayers using the mean-field approximation for temperatures close to T_m would at best be qualitative. No attempt was made here to make an elaborate fitting of model parameters to bring the $q = 10$ state model (within the Monte Carlo solution scheme) in close quantitative agreement with experimental data. However, the considerations presented by Mouritsen et al. (1983b) indicate that in principle this may be done. Thus, we conclude that the q-state models contain the essential physics for describing the gel-fluid phase transition of pure lipid bilayers provided that the value of q is chosen properly.

5.1.9 Computer simulation of the phase behavior of lipid bilayers with »impurities«: cholesterol, proteins, and polypeptides

From a biological point of view, the one-component lipid bilayer studied in the preceeding section is not particularly interesting. However, to make progress in the physical description of more realistic membranes it is an important first step to provide a reliable description of the background lipid bilayer matrix itself. Having accomplished this, the physicist may turn to the more complicated situation of introducing »impurities« in the membrane to simulate more realistic membranes consisting of several components.

An important membrane component is *cholesterol* which is an amphiphilic molecule with a rigid hydrophobically »smooth« sterol skeleton. Cholesterol, which enters as a lipid molecule in the bilayer, plays a crucial role as a regulator of membrane fluidity, and one of its potential effects on the life conditions of cells is well known in connection with *atheroschlerosis*. The first computer simulation study of the equilibrium properties of lipid-cholesterol mixtures is that performed by Mouritsen et al. (1983b). (An earlier computer study by Snyder and Freire (1980) of the lateral distribution of cholesterol in lipid bilayers has recently been shown not to provide the equilibrium properties of the mixture, but rather to be relevant to the kinetics of aggregation (Jan et al. 1984)).

There are a number of properties which a good model of a lipid bilayer containing cholesterol must account for. Three of the static properties are (i) the shape of the specific heat curves, (ii) the dependence of the transition enthalpy on cholesterol concentration c, and (iii) the existence and location of phase boundaries. The objective of the Monte Carlo study carried out by Mouritsen et al. (1983b) is to clarify whether a $q = 2$ state model can adequately describe the thermodynamic behavior of lipid-cholesterol mixtures or whether intermediate states have to be introduced to account for the experimental data. The $q = 2$ state model employed in this study is derived from the $q = 10$ state model by replacing the nine low-energy states by one temperature-dependent average »ground« state (Pink et al. 1980a). Each site of the triangular lattice is allowed to be occupied either by a lipid chain or a cholesterol molecule. Lipid chains cannot interchange positions directly on the lattice, but the cholesterol molecules are allowed to diffuse laterally by jumping from site to site. The specific mechanism of equilibrating the system is the following combination of Glauber and Kawasaki dynamics: A cholesterol molecule and one of its six nearest-neighbor sites are chosen at random. If the neighbor site is occupied by either another cholesterol or a lipid chain in its effective ground state, the cholesterol under consideration is not moved. If the site, however, is occupied by a lipid chain in its excited state, the cholesterol molecule is made to exchange places with the lipid chain. Two excitations per lipid chain are attempted between the procedures of offering all cholesterol molecules to move. This is in order that the lipid chains relax between each cholesterol movement. A lattice of $N = 900$ sites is studied with cholesterol concentrations ranging from $c = 0$ to $c = 0.5$. The lipid chains are taken to represent *DPPC* with the parameters resulting from the study of the pure lipid bilayer. Only nearest-neighbor interactions are considered and the interactions between two cholesterols and between a cholesterol and a lipid chain in its excited state are set to be zero. A trial value of $-0.4J_o$ is taken to represent the interaction between a cholesterol and a lipid chain in its ground state (the interaction between two lipid chains in their ground states is $-J_o$). Thus, cholesterol molecules prefer to have ordered chains adjacent to them. This value is chosen so that the temperature of the peak in the specific heat does not vary much with c. The system is started at 50 °C, and is equilibrated for between 50 and 100 Monte Carlo steps per site, then about 500 Monte Carlo steps per site are performed during which thermodynamic quantities are calculated. At the end of this procedure, the system is cooled by 2 °C and the procedure is repeated until 30 °C is reached. We calculate the enthalpy and the number of cholesterol- cholesterol nearest neighbors and second-nearest neighbors. It is found that the latter do not show any change over the range of temperatures studied, indicating that the system is not displaying pronounced phase separation. Figure 5.1.7 shows the calculated transition enthalpy, ΔH, as a function of c. ΔH is determined as the area under the specific heat peak. Two features are apparent. Firstly, the shape is similar to that measured in recent experiments (Maybrey et al. 1978), and predicted by Pink and Carroll (1978). It exhibits a convex shape at low c, followed by a tail at high values of c. In so far as this is concerned, the model is satisfactory. Secondly, the concentration at which ΔH shows the maximum slope occurs at $c \approx 0.25$. This is in contradiction to what is observed in experiments, where the maximum slope occurs between $c = 0.15$ and $c = 0.20$. The calculated transition enthalpy therefore persists to higher concentrations than is observed experimentally. We have tried other values of the interaction between cholesterol and lipid chains in their effective ground states and always found the same result. These interactions varied from $-0.3J_o$ to $-0.55J_o$. Values outside this range show that the position of the specific heat peak varies with concentration, which is not in accordance with experimental observations.

It is clear from the computer simulation results that a simple model of the kind employed in this study is inadequate to describe the equilibrium properties of lipid-

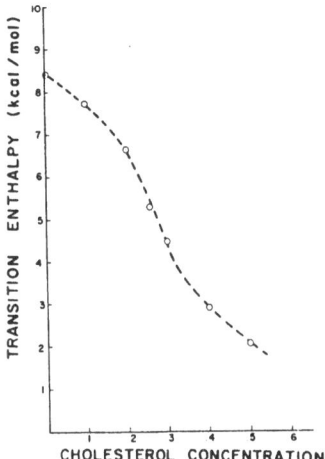

Fig. 5.1.7. Transition enthalpy as a function of cholesterol concentration in *DPPC* lipid bilayers. The Monte Carlo data (circles) are obtained for a $q = 2$ state model.

cholesterol bilayers. This may indicate either (i) different values of interaction constants must be chosen or (ii) the assumption, that only two lipid chain states are adequate to describe lipid-cholesterol bilayers, may be incorrect. A computer simulation study of a model which includes an intermediate state with jog and kink defects will be useful to provide guidance in constructing a more realistic model of lipid-cholesterol mixtures.

A major obstacle to the formulation of a theoretical model of a lipid bilayer with intrinsic proteins (or polypeptides) has been the very limited knowledge of the structure of membrane-bound proteins and thus the nature of the lipid-protein interaction. From the standpoint of statistical mechanics, the influence of proteins on the phase behavior and thermodynamic properties of the lipid matrix has been studied along two different routes. One is by means of phenomenological Landau theories (for a list of references and a description of the individual theories and their interrelations, see Mouritsen and Bloom 1984). The other route has involved the use of extended versions of microscopic interaction Hamiltonians originally developed to describe the pure lipid bilayer (Caillé et al. 1980). The approach chosen by Pink and coworkers (Pink and Chapman 1979, Tessier-Lavigne et al. 1982, Lookman and Pink 1984) follows the latter route and is built on the basis of the $q = 10$ state model for the pure lipid bilayer. The extended models include specific lipid-protein interactions but neglect direct protein-protein interactions. The only model incorporating proteins which so far has been studied by computer simulation (Lookman et al. 1983, Pink et al. 1983) is the $q = 2$ reduction (with the temperature-dependent ground state) of the $q = 10$ state model. The model accounts for two lipid-protein interaction constants, one for the interaction between the protein and a lipid chain in its ground state and one for the interaction between the protein and a lipid chain in its melted state. The two interaction constants are determined from a fit of the mean-field solution of the model to the concentration dependence of ^2H-NMR order parameters. This procedure, as well as the use of mean-field parameters in the pure lipid model, is questionable if one wants to make a quantitative comparison with experimental results on specific lipid-protein mixtures. In the Monte Carlo study performed by Pink et al. (1982, Lookman and Pink 1984) of this model, a scheme was devised to bring about the equilibrium of the mixture and simultaneously determine the lateral diffusion coefficient of various proteins and polypeptides. The proteins, which are represented by smooth hexagonal objects embedded in the triangular lipid lattice, are able to move from site to site by squeezing the lipids around the hexagons. The phase of the system is determined from the protein-protein correlation functions. For two different protein sizes, corresponding to molar weights of 5000 and 13500 of the hydrophobic segments, massive phase separation is encountered below the pure lipid melting point. The proteins are found to be almost completely immiscible in the lipid gel phase. The general shape of the phase diagram is in accordance with a variety of experiments (Chapman et al. 1979).

5.1.10 Have computer studies provided any new insight into the properties of biological membranes?

To answer this question at this very early stage of development of a theory for a biological membrane, it is conducive to recall the propagation of the physical experimental programs on membrane systems. While during the early seventies many experimentalists who were potentially interested in biology moved rather rapidly from physico-chemical systems related to membranes, such as soap solutions and liquid crystals, via lipid bilayers, to complex natural and reconstituted membranes, the last few years have witnessed a return to simpler and well-defined systems, such as one-component lipid bilayers with a single other component present. The main reason for this return has been the tremendous difficulties encountered when seeking an interpretation of experimental measurements on complex systems in terms of well-understood physical principles. However, the development up to the turning point has clearly demonstrated that simple lipid bilayers (with single species of cholesterol, polypeptides, or proteins present) are unique experimental and theoretical model systems for biological membranes. From the physicist's point of view, it is therefore of seminal importance to understand these simpler systems before there is any hope whatsoever of describing the structure and organization of biomembranes and from there, ultimately, approach the goal of relating structure and physiological function of membranes. In this sense, the physicist's approach is complementary to that of the traditional biochemist and membrane physiologist.

These considerations, then, bring us back to the original question. In the present author's opinion one may indeed gain valuable insight into the properties of biological membranes from theoretical studies of model membranes. In particular, computer studies of these systems are valuable since they at present constitute the only reliable way of calculating the properties of realistic molecular models. So far, the most comprehensive studies have been on models of pure lipid bilayers. The modelling of lipid-protein interactions is still in a very primitive stage and the microscopic models, which have been proposed so far, only seek to model the very gross features of the proteins. However, computer studies on such simplified models are extremely useful to determine some of the relevant variables inherent in the problem. Evidently, a major drawback is the large number of model parameters which have to be introduced even in a pure lipid bilayer model. Moreover, it is difficult to design experiments which can critically determine the model parameters.

Important topics to be adressed in future computer studies will among others be: (i) construction of a more realistic model of cholesterol in lipid bilayers, (ii) planar protein segregation in lipid bilayers, (iii) binary lipid mixtures, (iv) alcohol and anaesthetics in bilayers, and (v) bilayer organization in specific ionic environments relevant for membrane fusion.

5.2 Nuclear Dipolar Magnetic Ordering and Phase Transitions

5.2.1 Nuclear dipolar magnetic ordering

It was a major experimental achievement in the field of nuclear magnetism when in 1969 the nuclear magnetic resonance group at Sacley made the first observation of nuclear spin ordering governed by dipolar interactions (Chapellier et al. 1969). Since then, nuclear dipolar magnetic ordering has been observed experimentally in $^{42}Ca^{19}F_2$ (Goldman et al. 1974, Goldman 1977), in $^7Li^{19}F$ (Cox et al. 1975), in $^7Li^1H$ (Roinel et al. 1978), and in $^{42}Ca(^{16}O^1H)_2$ (Sprenkels et al. 1983). Due to the weakness of the magnetic dipole interactions, phase transitions and ordering phenomena typically take place in the range of $10^{-7} - 10^{-6}$K! In CaF_2, LiF, and $Ca(OH)_2$, the ordering is observed indirectly by nuclear magnetic resonance techniques. In LiH, the large pseudo-magnetic moments of the ^1H nuclei allow a direct investigation of the spin structures by means of neutron diffraction measurements.

The extremely low temperatures required for the creation of nuclear dipolar magnetic ordering are not accessible by any cryogenic technique currently available. Therefore, in the experiments only the nuclear spin degrees of freedom are cooled to the microdegree region. After cooling the entire system to about 0.3K by conventional ^3He refrigeration, the following two-step cooling process is employed (see e.g. Goldman 1977): (i) The nuclear spins are polarized in a high external field, $\bar{B}_0$, by coupling, via a microwave radiation, the nuclear spins to appropriate electronic spins introduced as a very dilute reservoir in the crystal. After this step, the Zeeman and the dipolar Hamiltonians become quasi-constants of motion and the nuclear spin system is described by a Zeeman spin temperature which is related to the nuclear spin polarization by an appropriate Langevin equation. In the second step, which only involves the nuclear spins, (ii) the Zeeman order is transformed adiabatically into dipolar order by introducing an interaction between the Zeeman and dipolar Hamiltonians. Instead of achieving this by lowering the static magnetic field $\bar{B}_0$, the crystal is exposed to a radio frequency field at a frequency close to the Larmor frequency of the nuclear spins. By sweeping the frequency slowly through resonance, the Zeeman system is demagnetized and the order is transformed adiabatically (i.e. by conserving the total entropy) into order governed by the dipolar Hamiltonian. The equilibrium distribution on the energy levels is then described by a dipolar spin temperature. If the initial polarization is sufficiently large (i.e. the entropy sufficiently low), the resulting dipolar spin temperature may enter the realm of spontaneous nuclear dipolar magnetic ordering.

After the two-step cooling process, the nuclear spin system is effectively decoupled from the lattice-degrees of freedom. Under these circumstances a spin system, which has both upper and lower bounds on its energy spectrum, can take on *positive* as well as *negative* spin temperatures. In the experimental step (ii) outlined above, the sign of the temperature is simply defined by the direction in which the resonance is approached.

At present, the major piece of information on nuclear dipolar magnetic ordering has been derived from nuclear magnetic resonance experiments (Abragam and Goldman 1982). Due to the local nature of these techniques, only indirect information on the nuclear spin ordering can be obtained, e.g. via the behavior of non-critical susceptibilities as a function of entropy (polarization) or dipolar energy. Entropy, rather than temperature, is the natural variable in this approach. To determine the temperature of the system, a theoretical model relating the entropy and the temperature has to be imposed. The direct structural studies of nuclear magnetism by neutron scattering (Roinel et al. 1978, 1980) are still in a very early stage, although an important breakthrough has established the feasibility of the approach for *LiH*.

5.2.2 The secular dipolar Hamiltonian

In the theoretical description of nuclear magnetic phenomena in diamagnetic crystals with light nuclei, the magnetic dipole-dipole interaction is an appropriate model (Abragam and Goldman 1982). Indirect couplings, such as the quadrupole coupling, are in these cases usually negligible.

As a consequence of the preparation of the spin system by the two-step cooling procedure described above, it is not the full dipole-dipole Hamiltonian, H_D, but rather the *secular dipolar Hamiltonian*, H'_D, (i.e. the part of H_D that commutes with the Zeeman Hamiltonian) which governs the ordering process. In the general case, for a system with two spin species — the I- and the S-spins —, the secular dipolar Hamiltonian is given by (Cox et al. 1975)

$$\frac{r_0^3}{\mu_I^2} H'_D = \sum_{j \in \omega_I} \sum_{k(\neq j) \in \omega_I} \sum_{p=x,y,z} A_p(\bar{r}_{jk}) I_{jp} I_{kp} + \sum_{j \in \omega_S} \sum_{k(\neq j) \in \omega_S} \sum_{p=x,y,z} B_p(\bar{r}_{jk}) S_{jp} S_{kp}$$

$$+ \sum_{j \in \omega_I} \sum_{k \in \omega_S} C_z(\bar{r}_{jk}) I_{jz} S_{kz} \tag{5.2.1}$$

where

$$A_z(\bar{r}_{jk}) = \tfrac{1}{2}(1 - 3\cos^2\theta_{jk})/r_{jk}^3$$

$$= -2A_x(\bar{r}_{jk}) \tag{5.2.2}$$

$$= -2A_y(\bar{r}_{jk})$$

$$B_z(\bar{r}_{jk}) = R^2 A_z(\bar{r}_{jk})$$
$$= -2 B_x(\bar{r}_{jk}) \tag{5.2.3}$$
$$= -2 B_y(\bar{r}_{jk})$$

$$C_z(\bar{r}_{jk}) = 2RA_z(\bar{r}_{jk}). \tag{5.2.4}$$

ω_I and ω_S denote the two spin reservoirs and $R = \mu_S/\mu_I$ is the ratio of the magnetic moments. θ_{jk} is the angle between the external static magnetic field $\bar{B}_0$, taken along the z-axis, and the vector $\bar{r}_{jk}$ connecting the jth and the kth lattice sites. $\bar{r}_{jk}$, which is time-independent, is in units of the lattice parameter r_0. It is noteworthy, that $\bar{B}_0$ does not enter the Hamiltonian directly but merely defines an axis of cylindrical symmetry.

For the systems to be considered here — CaF_2, LiF, and LiH — the magnetic nuclei are ^{19}F, ^1H, and ^7Li with spin quantum numbers $\frac{1}{2}$, $\frac{1}{2}$, and $\frac{3}{2}$, respectively. In these cases, the lattice formed by the magnetic nuclei is simple cubic, and the two spin reservoirs ω_I and ω_S constitute two interpenetrating face-centered cubic lattices. In the special case of $I = S$, which applies to the ^{19}F spins of CaF_2, the secular dipolar Hamiltonian is simplified by the constraint $B_{jk} = C_{jk} = 0$.

The theoretical work on the secular dipolar Hamiltonian (see Abragam and Goldman 1982 for a review) has predominantly sought to meet the need for a framework within which results of nuclear magnetic resonance experiments might be interpreted. The ordered structures have been predicted by the mean-field theory and by the Luttinger-Tisza method, and the temperature dependence of various quantities, such as entropy and bulk susceptibilities, has been derived from mean-field, spin-wave, and high-temperature restricted-trace calculations.

5.2.3 Perspectives in studies of nuclear dipolar magnetic ordering

The experimental feasibility of the production and observation of nuclear magnetic ordering has opened up an entirely new dimension in the study of cooperative magnetic phenomena. The unique features of this dimension manifest themselves when compared with features characteristic of conventional electronic magnetism:

i) The secular dipolar Hamiltonian governing the nuclear dipolar magnetic ordering is a »clean« model in that it contains no unknown or adjustable parameters. This is in contrast to electronic exchange Hamiltonians whose exchange integrals are very seldom known and which have been determined in the most favorable cases from a fitting of theories to experiment. Thus, the secular dipolar model offers a unique case for testing theories of cooperative phenomena.

ii) The equilibrium temperature can assume *positive* as well as *negative* values, in contrast to electronic spin systems in which the strong coupling of the spins to the lattice-degrees of freedom precludes negative temperatures. Hence, nuclear spin systems allow a probing of very fascinating aspects of general thermodynamics (Ramsey 1956).

iii) The long-range secular dipolar Hamiltonian implies a spatially inhomogeneous coupling distribution, in contrast to most short-range exchange interactions. Therefore, very complicated magnetic structures are expected.

iv) The coupling distribution of the secular dipolar Hamiltonian can be varied by changing the symmetry axis defined by the direction of $\overline{B}_o$. This is likely to result in a very rich phase behavior encompassing a variety of different spin structures and phase transitions between ordered and disordered phases as well as between ordered phases of different symmetry.

A number of factors to some extent outbalance the remarkable perspectives offered in the study of nuclear dipolar magnetic ordering. These are on the experimental side the extreme difficulties associated with the production and observation of the ordered states (Goldman et al. 1974). The accuracy and the details of the measurements are far from comparable with the precise information which can currently be obtained from electronic spin systems. In particular, scattering experiments on structural properties of nuclear spin systems are just emerging, and in the nuclear magnetic resonance studies only indirect information is obtained about the spin ordering. On the theoretical side, a treatment of the dipolar coupling is hampered by the complexity of the Hamiltonian, in particular by the long-range nature of the interaction.

5.2.4 Motivation for a numerical simulation study of nuclear dipolar magnetic ordering

Systems governed by the secular dipolar Hamiltonian, Eqs. (5.2.1) – (5.2.4), seem to be ideal objects for a computer simulation study classified as a *numerical experiment* (class (iii) in Chap. 1).

First of all, the model is »*clean*« with no unknown parameters. Thus, a numerical simulation of the model may be considered an experiment on an extremely realistic model system. Secondly, a numerical simulation can easily bypass the most serious experimental difficulties. The ordered states may be directly observed, the sign of the temperature may be chosen at will, and the temperature can in principle be made as low as required. Furthermore, very detailed information on the thermal and structural properties of the nuclear spin ordering can be obtained, e.g. through spatially displaced spin-spin correlation functions, order parameters, and a variety of response functions. In particular, critical fluctuations, such as ordering susceptibilities, and details about the location and the nature of phase transitions become available. Finally, a numerical study is free of any disturbing impurities which are likely to take part in the ordering. These impurities are first of all the paramagnetic centers introduced to facilitate the nuclear polarization, and secondly various spin-carrying isotopes in natural abundance, such as ^{43}Ca and ^{6}Li.

In comparison to conventional theoretical methods, computer simulation techniques are easily applied to a complicated Hamiltonian, such as Eqs. (5.2.1) – (5.2.4). The direction of $\overline{B}_0$ can be changed without introducing additional approximations. Most important, the computer simulation allows for thermal spin-

spin fluctuations which are completely neglected in the mean-field-like calculations. Furthermore, the same approach can be applied for all temperatures. The
latter is a distinct advantage over the approach chosen by Goldman and coworkers
who linked low-temperature mean-field calculations to high-temperature expansions with a highly unsatisfactory result in the transition region (Goldman et al.
1974).

Still, a numerical simulation study of the secular dipolar Hamiltonian is not
without its problems. There are basically two major difficulties, namely the long-
range character of the dipolar coupling and the quantum nature of the nuclear
spins. In the numerical simulations reported below, the Hamiltonian is simplified
to a tractable form by truncating the range of interaction at some finite distance
and by imposing a classical approximation for the spins. The truncation of the
range of interaction may in some cases influence the ordered spin structures. Local
quantities such as the energy and the non-ordering susceptibilities will not seriously
depend on the truncation, provided that the magnetic structure is unaltered. In
contrast, the finer details of critical behavior, such as critical exponents, might
well depend on whether the range of interaction is finite or not. The classical
approximation for the spins is expected to be immaterial regarding the spin structures and the universal critical properties. However, absolute values of transition
temperatures, energies, and susceptibilities obtained for the classical system should
not be directly compared with those of the quantum systems. The interpretation of
most of the experimental results on nuclear dipolar ordering has so far relied very
heavily on mean-field-like theoretical calculations. The confidence to be placed
in this mean-field approach may very well be evaluated by comparing mean-field
and computer simulation results for the same truncated classical secular dipolar
Hamiltonian, despite the fact that this simplified model Hamiltonian on many accounts is expected to have different properties from the long-range secular dipolar
quantum Hamiltonian.

5.2.5 Monte Carlo studies of systems with truncated classical secular dipolar interactions

In a series of papers, Mouritsen and Knak Jensen (1978, 1980a, 1981) have studied
aspects of nuclear dipolar magnetic ordering utilizing Monte Carlo techniques. For
systems with a single spin species, such as CaF_2, six different cross-sections of the
phase diagram have been studied, corresponding to three different directions of $\overline{B}_o$
at positive as well as at negative spin temperatures. The three orientations of $\overline{B}_o$ are
those conventionally considered in experiments, i.e. $\overline{B}_o$ along the directions [001],
[110], and [111] of the simple cubic lattice formed by the ^{19}F spins. $\overline{B}_o$ breaks the
cubic symmetry differently in the three cases. For $\overline{B}_o \mid [001]$, the system still has a
four-fold axis, whereas for $\overline{B}_o \mid [110]$ and $\overline{B}_o \mid [111]$ the highest symmetry elements
are a two-fold and a three-fold axis, respectively. The range of interaction is
usually truncated after the 26 nearest neighbors. Occasionally, an interaction cube
involving 124 spins has been used. The regimes of direct interaction are chosen so
as to preserve the symmetry of H'_D.

The computer simulations are carried out on a number of different lattice sizes, $N = L^3$, with L ranging from 6 to 20. The lattices are subject to toroidal periodic boundary conditions. The classical spins are represented by unit vectors, $\bar{e}_j = (e_{xj}, e_{yj}, e_{zj})$.

5.2.6   Nature of the spin structures: »Permanent« structures or the devil's staircase?

The static spin structures at »low« temperatures, $T \simeq \pm 0$, have been inferred from direct inspection of spin configurations and from spin-spin correlation functions calculated at low values of $|T|$, typically $0.1\mu^2/r_o^3 k_B$. Special precautions have to be taken in order to avoid metastable states and incommensurability of the spin structure with the lattice size. Thus, the Monte Carlo calculations are performed for several different initial disordered configurations and for various lattice sizes. It turns out that when the criterion for commensurability is fulfilled, it is fairly easy to anneal the systems and produce a single domain of ordering. The fast domain-growth kinetics following quenches to low temperatures is presumably caused by the continuous nature of the spins and the fact that the order parameter dimensionalities are found to be less or equal to the spatial dimension (cf. Sec. 5.4). In all cases, two-sublattice antiferromagnetic spin structures are found. These are listed below in terms of the star of the propagation vector, $\bar{q}$, together with the dimension, n, of the order parameter. Following Landau theory, n is defined as the dimension of the irreducible representation of the symmetry group of the paramagnetic phase according to which the order parameter transforms. n is given by $n = sl_p$ (Mukamel and Krinsky 1976a), where s is the dimension of the star of the propagation vector, $\bar{q}$, and l_p is the dimension of the small representation.

Systems with a single spin species: CaF$_2$
(a) $\bar{B}_o \,|\, [001]$ $T = +0$

$$\bar{e}_j = (\cos a, \sin a, 0) \cos(\bar{q} \cdot \bar{r}_j), \quad |a| \le \pi$$

$$\bar{q} = \frac{2\pi}{r_o}(0, 0, \tfrac{1}{2}) \tag{5.2.5}$$

$$n = 2.$$

This structure is accidentally degenerate with

$$\bar{e}_j = (0, 0, 1) \cos(\bar{q}_i \cdot \bar{r}_j)$$

$$\bar{q}_1 = \frac{2\pi}{r_o}(\tfrac{1}{2}, 0, 0),$$

$$\bar{q}_2 = \frac{2\pi}{r_o}(0, \tfrac{1}{2}, 0) \tag{5.2.6}$$

$$n = 2.$$

(b) $\bar{B}_o \mid [001]\ T = -0$

$$\bar{e}_j = (0, 0, 1) \cos(\bar{q} \cdot \bar{r}_j)$$

$$\bar{q} = \frac{2\pi}{r_o}(0, 0, \tfrac{1}{2}) \tag{5.2.7}$$

$$n = 1.$$

(c) $\bar{B}_o \mid [110]\ T = +0$

$$\bar{e}_j = (1, 1, 0) \cos(\bar{q} \cdot \bar{r}_j)/\sqrt{2}$$

$$\bar{q} = \frac{2\pi}{r_o}(0, 0, \tfrac{1}{2}) \tag{5.2.8}$$

$$n = 1.$$

(d) $\bar{B}_o \mid [110]\ T = -0$

$$\bar{e}_j = (1, 1, 0) \cos(\bar{q} \cdot \bar{r}_j + \tfrac{\pi}{4})$$

$$\bar{q} = \frac{2\pi}{r_o}(\tfrac{1}{4}, \tfrac{1}{4}, 0) \tag{5.2.9}$$

$$n = 1.$$

(e) $\bar{B}_o \mid [111]\ T = +0$

$$\bar{e}_j = (1, 1, 1) \cos(\bar{q}_i \cdot \bar{r}_j + \tfrac{\pi}{4})\sqrt{2}/\sqrt{3}$$

$$\bar{q}_1 = \frac{2\pi}{r_o}(\tfrac{1}{4}, -\tfrac{1}{4}, 0),$$

$$\bar{q}_2 = \frac{2\pi}{r_o}(\tfrac{1}{4}, 0, -\tfrac{1}{4}), \tag{5.2.10}$$

$$\bar{q}_3 = \frac{2\pi}{r_o}(0, \tfrac{1}{4}, -\tfrac{1}{4})$$

$$n = 3.$$

(f) $\bar{B}_o \mid [111]\ T = -0$

$$\bar{e}_j = (1, 1, 1) \cos(\bar{q} \cdot \bar{r}_j + \tfrac{\pi}{4})\sqrt{2}/\sqrt{3}$$

$$\bar{q} = \frac{2\pi}{r_o}(\tfrac{1}{4}, \tfrac{1}{4}, \tfrac{1}{4}) \tag{5.2.11}$$

$$n = 1.$$

These structures, whose derivation is conditioned by an interaction cube involving 26 nearest neighbors, fall naturally into two classes. The first class, (a) – (c), consists of *double-layered structures* with layer magnetizations of the sign sequence $+ - + - \ldots$ along the appropriate propagation vector. The second class, (d) - (f), is made up of *multi-layered structures* with the sign sequence $+ + - - + + \ldots$ of the layer magnetizations. Since this classification also turns out to be relevant with respect to the phase behavior of the systems, the computer simulation results for the phase transitions of these two classes will be presented separately in Secs. 5.2.7 and 5.2.8.

Systems with two spin species: LiF and LiH

Here only the case $\overline{B}_o \mid [001]$ $T = -0$ has been investigated by computer simulation (Mouritsen and Knak Jensen 1980a). The structure is found to be identical to the one given in Eq. (5.2.7) for all positive values of the ratio of magnetic moments (values in the range $0 \le R \le 10$ have been investigated). Thus, each sublattice contains both I and S spins, and for $R \ne 1$ the structure may be termed antiferrimagnetic.

The structures listed above are the $T = \pm 0$ ground states. It is not *a priori* clear that these structures are »permanent« in the sense that they stay stable all the way up to the transition point. First of all, it should be noted that the degeneracy at $T = +0$ of the structures in Eqs. (5.2.5) and (5.2.6) is lifted at any finite temperature since the longitudinal structure in Eq. (5.2.6) is unstable for entropy reasons. This is also borne out by the Monte Carlo calculations and by series analysis of the ordering susceptibility (Mouritsen 1979). One approximate way of determining the »permanency« of a structure is to compare the ground states with the structures predicted by mean-field theory to be stable in the immediate vicinity of the phase transition. Assuming the transition to be continuous, mean-field theory predicts the stable structures to be described by the propagation vectors at the extrema of the Fourier transform, $A_p(\overline{q})$, of the coupling distribution (Villain 1959)

$$A_p(\overline{q}) = \sum_k A_p(\overline{r}_{jk}) e^{i\overline{q} \cdot \overline{r}_{jk}}. \tag{5.2.12}$$

The minimum and maximum of Eq (5.2.12) determine the stable structures at positive and negative temperatures, respectively. With $A_p(\overline{r}_{jk})$ given by Eq. (5.2.2), we recover from Eq. (5.2.12) the magnetic structures in Eqs. (5.2.5), (5.2.7), (5.2.8), and (5.2.10) for the cases (a), (b), (c), and (e). The »permanency« of these structures is confirmed by finite-temperature Monte Carlo calculations. A similar result is found for the system with two spin species for $\overline{B}_o \mid [001]$ and $T < 0$.

However, in the cases (d) and (f) the stable spin structures resulting from Eq. (5.2.12), though still longitudinal, are found to be described by the propagation vectors $\overline{q} = \frac{2\pi}{r_o} q^* (1, 1, 0)$ and $\overline{q} = \frac{2\pi}{r_o} q^* (1, 1, 1)$, with $q^* = 0.2736$ and $q^* = 0.2051$, respectively. Since the periodicity is different at $|T| = 0$ and at the transition, the wave vector, $\overline{q}(T)$, of the modulation is expected to vary with temperature. The basic physical reason for this peculiarity is the presence of competing interactions in the Hamiltonian. The question now naturally arises: What is the precise form of $\overline{q}(T)$ and does that form have any physical significance?

This question is closely related to whether a devil's staircase exists in modulated systems with competing periodicities (see e.g. Bak 1982 for a review). In view of the considerable theoretical interest in these problems, we shall discuss in detail the possibility of a devil's-staircase behavior in systems with truncated secular dipolar interactions.

To this end, we have performed mean-field and Monte Carlo finite-temperature calculations in the cases (d) and (f) (Mouritsen and Knak Jensen 1981). Both types of calculations are performed on a number of finite periodic lattices of liniear dimension L. It should be emphasized that finite periodic lattices can only accomodate stable phases which are commensurate with L, i.e. only structures with $q = n/L, n = 1, 2, \ldots, L$, are allowed (see also Selke and Fisher 1979). Even if the true structure was actually incommensurate, it would lock-in at a nearby commensurate value of q. The results of the mean-field calculations for both cases (d) and (f) indicate that a number of rational values of q between $q = \frac{1}{4}$ and $q = q^*$ become stable. However, not necessarily all of the possibilities occur. For example, for $\overline{B}_o \mid [111]$ $T < 0$ the structures with $q = \frac{5}{20}$ and $\frac{4}{20}$ are stable for $L = 20$ whereas the structures with $q = \frac{25}{100}, \frac{22}{100}$, and $\frac{21}{100}$ are stable for $L = 100$. The general trend of the mean-field calculation is that an increase of L leads to more stable phases with wave vectors in the interval $[\frac{1}{4}, q^*]$. These results therefore suggest that a staircase behavior of the wave vector $q(T)$ should be expected in the cases (d) and (f). It is impossible from these calculations to assess whether this staircase is a real devil's staircase or a »harmless« staircase with only a finite number of steps (Villain and Gordon 1980). We then turn to the Monte Carlo results. For $\overline{B}_o \mid [111]$ $T < 0$, simulations have been performed on lattices with $L = 12$ and 20. In the smaller lattice, the square-wave structure with $q = \frac{1}{4}$ remains stable up to the transition to the disordered phase. For the larger lattice, a sinusoidally modulated phase with $q = \frac{1}{5}$ intervenes. Both these results are in accordance with mean-field theory. The numerical determination of the harmonic content of a structure proceeds via a Fourier analysis of the z-component of the coarse-grained averages of the layer magnetizations. As an example, Fig. 5.2.1 shows a plot of the order parameter amplitude, $m(q) = (a_q^2 + b_q^2)^{\frac{1}{2}}$, where a_q and b_q are the cosine and sine Fourier components. The plot in this figure corresponds to a temperature at which a transition between the two ordered structures occurs in a lattice with $L = 20$. For $\overline{B}_o \mid [110]$ $T < 0$, Monte Carlo calculations on lattices with $L = 8, 12$, and 16 only lead to one stable ordered structure, $q = \frac{1}{4}$. This is again in accordance with mean-field theory which predicts that much larger lattices are required to accomodate intermediate phases since q is predicted to lie in the interval $[\frac{1}{4}, q^* = 0.2736]$. It is impractical to do computer simulations for a complicated Hamiltonian, such as the secular dipolar Hamiltonian, on lattices very much larger than 20^3. We are thus unable to persue this fascinating scenario of multiple phase changes (simpler models, such as the *ANNNI* models are more suitable for a systematic study, see e.g. Selke and Fisher 1979, Rasmussen and Knak Jensen 1981). However, the concordant trends in the Monte Carlo and mean-field calculations lead us to the conjecture that the phase behavior in the cases $\overline{B}_o \mid [110]$ $T < 0$ and $\overline{B}_o \mid [111]$ $T < 0$ is staircase-like, in the sense that more and more phases become stable when the calculations become more and more refined.

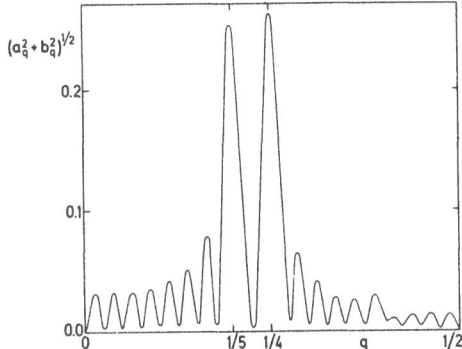

Fig. 5.2.1. Fourier spectrum of the layer magnetizations for $\overline{B}_o \,|\, [111]\, T < 0$ at a temperature where a lattice with 20^3 spins undergoes a first-order phase transition between two ordered structures characterized by wave vectors $q = \frac{1}{4}$ and $q = \frac{1}{5}$.

We now proceed to compare the spin structures obtained for the truncated secular dipolar Hamiltonian with those derived theoretically and those found experimentally for the truly long-range coupling. (Abragam and Goldman 1982). So far, experimental observation of ordered structures has been successful for CaF_2 in the cases $\overline{B}_o \,|\, [001]\, T > 0$, $T < 0$, $\overline{B}_o \,|\, [110]\, T > 0$, $T < 0$, and $\overline{B}_o \,|\, [111]\, T < 0$; for LiF in the case $\overline{B}_o \,|\, [001]\, T < 0$; and for LiH in the cases $\overline{B}_o \,|\, [001]\, T < 0$, $T > 0$, and $\overline{B}_o \,|\, [110]\, T > 0$, $T < 0$. The theoretical predictions for the long-range coupling are in all cases in accordance with the experimental findings. The structures are found to be the same for the one- and two-spin species systems provided that the experimental conditions are identical. This is due to the fact that the values of R involved (0.86 for LiH and 0.81 for LiF) are close to unity. It thus suffices to discuss the general six experimental situations specified by the direction of $\overline{B}_o$ and by the sign of the temperature. For the double-layered structures, (a) – (c), the predictions for the long- and short-range Hamiltonians are the same, except for $\overline{B}_o \,|\, [001]\, T > 0$ where the experiments suggest a longitudinal structure

(a′) $\quad \overline{B}_o \,|\, [001]\, T = +0$

$$\overline{e}_j = (0, 0, 1) \cos(\overline{q} \cdot \overline{r}_j)$$

$$\overline{q} = \frac{2\pi}{r_o} (\tfrac{1}{2}, \tfrac{1}{2}, 0) \tag{5.2.13}$$

$$n = 1.$$

This discrepancy is easily removed by enlarging the interaction cube of the truncated interaction to include 124 neighbors, in which case the structure in Eq. (5.2.13) is obtained. As far as the multi-layered cases, (d) – (f), are concerned nuclear magnetic resonance as well as neutron diffraction experiments for $\overline{B}_o \,|\, [110]\, T < 0$ and $\overline{B}_o \,|\, [111]\, T > 0$ indicate that the ordered states may be characterized as »ferrosandwiches,« i.e. ferromagnetic states with alternating domains perpendicular to the direction of $\overline{B}_o$. The neutron experiments (Roinel et al. 1980) show that the distance between the domain walls are comparable to typical distances between the paramagnetic impurities, indicating that the domain structure is not necessarily caused by the dipolar coupling itself. It is noteworthy that the modulation found in the experiments follows the same direction as that of the staircase phases for the truncated model. Thus, it is interesting to speculate whether the experimental systems may indeed exhibit a staircase behavior with certain periodicities being locked-in by the impurities. To clarify these problems, it is nessary to do experiments which first of all bring the system further down in temperature than just beyond the transition point and which secondly use a more dilute reservoir of electronic spins. Since these two requirements are to some extent in conflict with each other, such an experimental program is expected to be difficult to carry out with present techniques.

5.2.7 Double-layered spin structures in CaF_2-like systems: Continuous transitions and critical behavior

In this section, we describe the Monte Carlo results for the static thermodynamic properties of the three situations $\overline{B}_o \mid [001] T > 0$, $T < 0$ and $\overline{B}_o \mid [110] T > 0$. Special attention will be paid to the phase transition which in all three cases is found to be continuous. It is beyond experimental reach at present to obtain any detailed information about the nature of the phase transition, not to speak of finer details, such as critical exponents. A computer simulation study is therefore desirable in order to obtain some additional experimental information. It should, however, be kept in mind that the critical properties of the truncated secular dipolar model are not necessarily those of the truly long-range dipolar model. Still, the truncated model merits study in its own right. In particular, it is of interest to relate its critical behavior to the universal classification of continuous phase transitions.

A great variety of static properties and their temperature dependence has been calculated, such as internal energy, order parameter, specific heat, ordering- and non-ordering susceptibilities, and spin-spin correlation functions (Mouritsen and Knak Jensen 1978). The continuous nature of the phase transition is established from the continuous variation of the internal energy and the order parameter throughout the transition region, irrespective of the lattice size. In Fig. 5.2.2 is shown the order parameter, m_1^s/m_o (the staggered magnetization), as a function of inverse temperature in the case $\overline{B}_o \mid [001] T < 0$. The various components of the bulk and staggered susceptibility tensors are shown in Figs. 5.2.3 and 5.2.4 for the case $\overline{B}_o \mid [110] T > 0$. The bulk susceptibilities in Fig. 5.2.3 are qualitatively similar to those observed experimentally for CaF_2. A quantitative comparison should not be attempted due to the difference in spin quantum numbers between the theoretical model and CaF_2. Figure 5.2.3 shows that $\chi_\perp^b(T)$ has a slight temperature dependence in the antiferromagnetic phase, in contrast to the mean-field theory which predicts $\chi_\perp^b(T)$ to be a constant. The ordering (staggered) susceptibility, $\chi_1^s(T)$, which unfortunately has not been studied experimentally as yet, is strongly peaked at the transition point. Similarly, the heat capacity is singular at the transition. From the position of these singularities, the transition

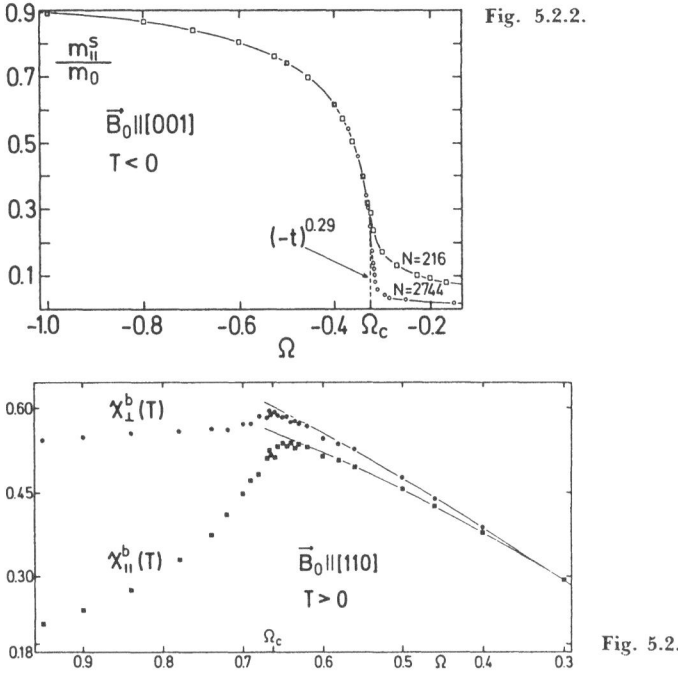

Fig. 5.2.2.

Fig. 5.2.3.

Fig. 5.2.2. Order parameter, $m_\parallel^s/m_o$, as a function of inverse temperature in the case $\bar{B}_o \mid [001]\ T < 0$. Monte Carlo data are given for two different lattice sizes N. The dashed line is the function $(-t)^{0.29}$ with $t = (T - T_c)/T_c$. $\Omega = \mu^2/r_o^3 k_B T$.

Fig. 5.2.3. Transverse, $\chi_\perp^b(T)$, and longitudinal, $\chi_\parallel^b(T)$, bulk susceptibility per spin for the case $\bar{B}_o \mid [110]\ T > 0$. Monte Carlo data are given for a system with $N = 12^3$ spins. The full lines are the sixth order high-temperature series expansions.

temperature of the finite system may be estimated. A full finite-size scaling analysis of the transition temperatures has not been carried out for this model, but the infinite-lattice critical temperature, T_c, may with a high degree of confidence be determined with 1% accuracy from the transition temperature of the larger lattices.

Having the critical temperature T_c at hand, we can proceed to determine critical exponents. Only the cases $\bar{B}_o \mid [001]T < 0$ and $\bar{B}_o \mid [110]\ T > 0$ have been investigated by Monte Carlo calculations in sufficient depth to allow a determination of critical exponents. We shall focus on the exponents γ and ν pertaining to the ordering susceptibility and correlation length, respectively. It suffices to briefly mention that the order parameter exponent β may also be extracted from the Monte Carlo data. In fact, it turns out that a simple power law $(-t)^\beta$, $t = (T - T_c)/T_c$, with the *same* value of β describes the order parameter data in both cases from about $t = -10^{-2}$ all the way down to zero temperature. The resulting value of β, $\beta \simeq 0.30 \pm 0.03$, is indistinguishable from the renormalization group value of $\beta_{RG} = 0.325$ for the three-dimensional spin-$\frac{1}{2}$ Ising model (Le Guillou and Zinn-Justin 1980).

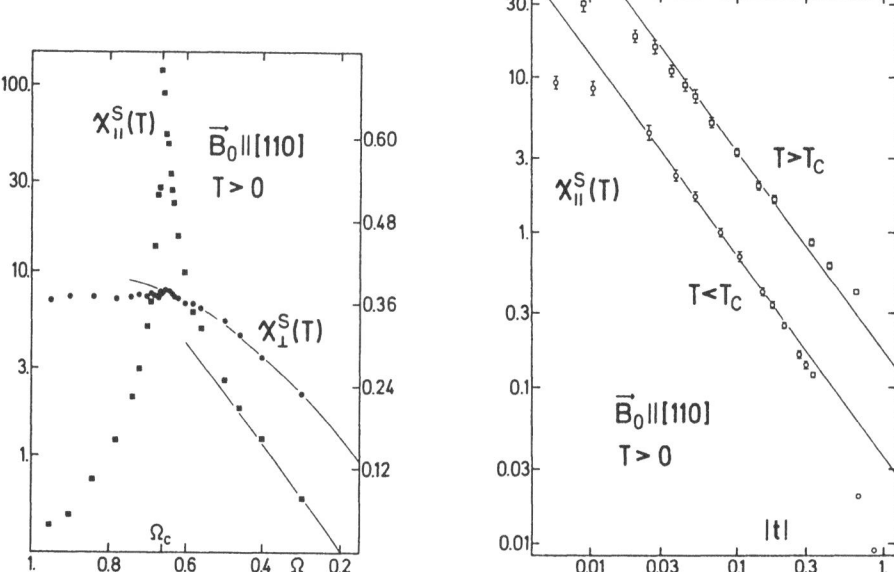

Fig. 5.2.4. Transverse, $\chi^s_\perp(T)$, and longitudinal, $\chi^s_{||}(T)$, staggered susceptibility per spin. The symbols are explained in Fig. 5.2.3. The logarithmic left-hand scale refers to $\chi^s_{||}(T)$, and the right-hand scale refers to $\chi^s_\perp(T)$.

Fig. 5.2.5. Log-log plot of the ordering susceptibility, $\chi^s_{||}(T)$, vs reduced temperature, t, in the case $\overline{B}_o \mid [110]\ T > 0$. The Monte Carlo data are obtained for a system with 12^3 spins.

Figure 5.2.5 shows a log-log plot of the ordering susceptibility, $\chi^s_{||}(T)$, vs reduced temperature, t, in the case $\overline{B}_o \mid [110]\ T > 0$. An analysis of the divergence of $\chi^s_{||}(T)$ in terms of the simple power-law singularities, $\chi^s_{||}(T) \simeq C_+ t^{-\gamma}$ for $t \to 0_+$ and $\chi^s_{||}(T) \simeq C_-(-t)^{-\gamma}$ for $t \to 0_-$, leads to values of the critical exponents as listed in Table 5.2.1. This table also contains the exponents for the case $\overline{B}_o \mid [001]\ T < 0$. The simple power-law expressions are found to be valid typically in the range $0.02 \le |t| \le 0.4$. Deviations from pure power-law behavior are found closer to the critical points due to finite-size rounding and further away from the critical points due to correction-to-scaling. Figure 5.2.6 is a log-log plot of the inverse correlation length $\kappa(T)$ vs reduced temperature in the case $\overline{B}_o \mid [001]\ T < 0$. The correlation length is extracted from a generalized Ornstein-Zernike ansatz for the spatial and thermal decay of the longitudinal spin-spin correlation function (Mouritsen and Knak Jensen 1978). For an anisotropic interaction, $\kappa(T)$ depends on the direction in which the decay is observed, and Fig. 5.2.6 therefore includes the correlation length parallel and perpendicular to the easy axis of the Hamiltonian. The figure demonstrates very clearly that the thermal decay of $\kappa(T)$ obeys simple power laws, $\kappa(T) \simeq F_+ t^\nu$ and $\kappa(T) \simeq F_-(-t)^{\nu'}$, over about a decade up from $|t| \simeq 0.015$. Moreover, the exponent ν is independent of direction. The critical exponents are listed in Table 5.2.1. It should be noted that the exponents obey the symmetric scaling relations, $\gamma = \gamma'$ and $\nu = \nu'$ (cf. Eq. (2.1.12)).

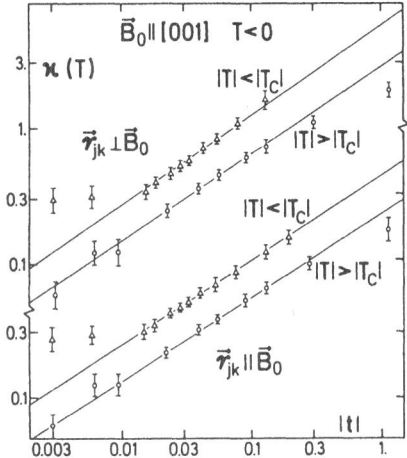

Fig. 5.2.6. Log-log plot of the inverse correlation length, $\kappa(T)$, vs reduced temperature, t, in the case $\overline{B}_o \mid [001]\ T < 0$. The thermal decay of $\kappa(T)$ is given for two different directions specified by $\overline{r}_{jk}$. The Monte Carlo data derive from a system with 20^3 spins.

Table 5.2.1 also includes the critical exponents γ, ν, and η as derived from high-temperature series analysis of the ordering susceptibility and of the second spherical moment of the longitudinal pair correlation functions (Mouritsen 1979). η is the exponent describing the spatial decay of the correlation function at the critical point, $\Gamma_1(\overline{r}, T = T_c) \sim |\overline{r}|^{-(1+\eta)}$. The various series and their combinations have been subjected to standard asymptotic analysis, combining ratio, Neville, dlog-Padé, and critical-point renormalization methods with appropriate conformal transformations isolating the physical singularities. The analyses have been carried out in terms of simple power-law divergences and confluent singularities have been neglected. It is obvious from Table 5.2.1 that there is an overall agreement between the exponents estimated by the two different approaches.

The critical parameters are now discussed in the context of universality. According to the universality hypothesis for static critical phenomena (cf. Sec. 2.1.2), only a few basic properties of a system are necessary to classify the transition and to determine the universal critical parameters. These basic properties are spatial dimension, dimensionality of the order parameter, and the range of interaction (Fisher 1974). In contrast, a number of details are irrelevant, such as spin quantum number and degree of anisotropy in the coupling distribution (including lattice structure). Consequently, the phase transitions in the cases $\overline{B}_o \mid [001]\ T < 0$ and $\overline{B}_o \mid [110]\ T > 0$ should be in the same universality class; they are both three dimensional and characterized by $n = 1$. This is indeed supported by the Monte Carlo results which for these two different situations lead to the same set of critical exponents (cf. Table 5.2.1). Among universal parameters are also predicted to be certain combinations of critical amplitudes, e.g. F_+/F_- and C_+/C_- (Bervillier 1976, Brézin et al. 1976). The Monte Carlo calculations predict a number of different values of the amplitudes $F_\pm$ depending on direction in the lattice relative

3-d system	Method	n
$\overline{B}_o \mid [001] \; T < 0$	Series[a] Monte Carlo[b]	1
$\overline{B}_o \mid [110] \; T > 0$	Series[a] Monte Carlo[b]	1
$\overline{B}_o \mid [111] \; T < 0$	Series $(q = \frac{1}{5})$[a]	1
Spin-$\frac{1}{2}$ Ising model	Series[c] Monte Carlo[d]	1
Landau-Ginzburg-Wilson	Second-order ϵ-expansion[e]	1
$(\phi^2)^2$-field theory	Perturbation expansion[f]	1
MnF_2	Neutron scattering[g]	1
Beta-brass	Neutron scattering[h]	1
Xe	Light scattering[i]	1
$\overline{B}_o \mid [001] \; T > 0$	Series[a]	2
3-d planar models	Series[j]	2
Landau-Ginzburg-Wilson	Second-order ϵ-expansion[e]	2
$(\phi^2)^2$-field theory	Perturbation expansion[f]	2
Superfluid ^4He	Superfluid density measurement[k]	2

Table 5.2.1. Critical exponents for various three-dimensional systems with short-range interactions. The table is divided into two sections corresponding to the two universality classes characterized by dimensionalities $n = 1$ and $n = 2$ of the order parameter. Exponents are given for real materials as well as for model systems studied by different theoretical methods. The primed exponents correspond to the ordered phase and the unprimed ones to the disordered phase. * (third order).

to $\overline{B}_o$. However, in all cases the *ratio* F_+/F_- is the same. The values of the ratios C_+/C_- and F_+/F_- are listed in Table 5.2.2 and are also seen to be consistent with universality.

A number of three-dimensional materials and models (notably the ferromagnetic Ising model) are also predicted to belong to the same universality class as the one studied here. Tables 5.2.1 and 5.2.2 contain some universal critical parameters for these systems as obtained from a variety of theoretical and experimental techniques. Although the Monte Carlo results for the truncated secular dipolar

γ'	γ	ν'	ν	η
1.32 ± 0.07	1.25 ± 0.02 1.27 ± 0.07	0.65 ± 0.06	0.63 ± 0.01 0.65 ± 0.06	0.05 ± 0.03
1.30 ± 0.05	1.24 ± 0.03 1.25 ± 0.05	0.61 ± 0.05	0.62 ± 0.02 0.61 ± 0.05	0.05 ± 0.05
	1.25 ± 0.03		0.62 ± 0.02	0.05 ± 0.05
1.29 ± 0.05	1.250 ± 0.002 1.23 ± 0.05	0.643 ± 0.010	0.638 ± 0.02 0.61 ± 0.05	0.04 ± 0.01
	1.244		0.627	0.037*
	1.240		0.630	0.032
	1.24 ± 0.02		0.634 ± 0.020	0.05 ± 0.02
	1.24 ± 0.015		0.65 ± 0.02	0.081 ± 0.072
	1.246 ± 0.010			
	1.33 ± 0.02		0.68 ± 0.01	0.09 ± 0.07
	1.333 ± 0.010		0.678 ± 0.005	0.031 ± 0.015
	1.300		0.650	0.034*
	1.316		0.669	0.034
			0.666 ± 0.006	

[a]Mouritsen 1979. [b]Mouritsen and Knak Jensen 1978. [c]Roskies 1981, Ferer and Velgakis 1983, Essam and Sykes 1963. [d]Mouritsen and Knak Jensen (unpublished). [e]Wilson and Kogut 1974. [f]Le Guillou and Zinn-Justin 1980. [g]Schulhof et al. 1970. [h]Als-Nielsen 1976. [i]Güttinger and Cannell 1981. [j]Rogiers et al. 1978, 1979. [k]Tyson and Douglass 1966.

	3-d Ising	$\overline{B}_o \mid [001]\, T < 0$	$\overline{B}_o \mid [110]\, T > 0$
C_+/C_-	5.06 ± 0.08[a]	6.8 ± 2.3	5.3 ± 1.5
F_+/F_-	0.513 ± 0.003[a]	0.50 ± 0.10	0.55 ± 0.12

Table 5.2.2. Universal ratios of critical amplitudes for models in the universality class of the three-dimensional Ising model. $F_\pm$ and $C_\pm$ are the amplitudes for the correlation length and the ordering susceptibility, respectively. [a]Fisher and Tarko 1975.

interaction do not have a quality comparable to what can be obtained from sophisticated experimental techniques or theoretical calculations on simple models, they are consistent with and support the concept of universality.

5.2.8 Multi-layered spin structures in CaF_2-like systems: First-order phase transitions

In the case $\overline{B}_o \mid [111]\ T > 0$, the triple degenerate structure, Eq. (5.2.10), stays stable throughout the ordered phase and eventually undergoes a first-order transition to the paramagnetic state. The first-order nature is signalled by discontinuities in the internal energy and in the order parameter at the transition as well as by the presence of hysteresis. This is demonstrated in Fig. 5.2.7 which should be contrasted to Fig. 5.2.2. A similar pronounced first-order transition is encountered in the case $\overline{B}_o \mid [110]\ T < 0$ for systems of linear dimensions $L = 8, 12$, and 16. As explained in Sec. 5.2.6, this finding may not be characteristic of the thermodynamic limit where a staircase behavior is expected from mean-field theory. The case $\overline{B}_o \mid [111]\ T < 0$ has been extensively described in Sec. 5.2.6. Here it suffices to mention that the transition in the large system, $L = 20$, between the two ordered phases, described by wave vectors $q = \frac{1}{4}$ and $q = \frac{1}{5}$, is of first order in agreement with the mean-field prediction. Very long-lived metastabilities are associated with this transition. The transition from the intermediate phase characterized by $q = \frac{1}{5}$ to the disordered phase may be continuous. No metastabilities have been encountered, and any discontinuities in internal energy and order parameter are below the resolution of the numerical simulation.

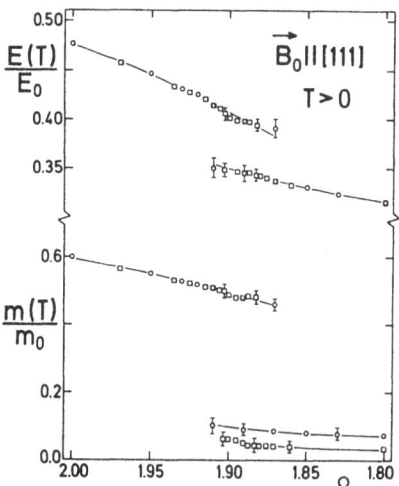

Fig. 5.2.7. Normalized internal energy, $E(T)/E_o$, and order parameter, $m(T)/m_o$, as functions of inverse temperature, $\Omega = \mu^2/r_o^3 k_B T$, in the case $\overline{B}_o \mid [111]\ T > 0$. The data are obtained from Monte Carlo calculations on systems with $N = 8^3$ (circles) and $N = 12^3$ (squares) spins.

5.2.9   Can series expansions provide information on the nature of the phase transitions?

High-temperature series analyses of various properties of a model system undergoing a phase transition can give no clear-cut answer concerning the order of the transition. To solve this problem, a combined study of high- and low-temperature series for the free energy may be more appropriate (see e.g. Plischke and Zobin 1977, Saul et al. 1974). A finite series reflects the behavior of the system some distance from criticality. Even if the system in this region behaves in a manner which gives rise to effective critical singularities of the extrapolated series, it is impossible to exclude that the system does not eventually undergo a first-order transition.

From a microscopic point of view, the six models studied in this chapter are very similar. Since the Monte Carlo calculations have predicted continuous as well as first-order transitions, this family of six models offers an exceptional case for investigating how high-temperature series analyses may be utilized to assess the order of a phase transition. The strategy to be followed is to subject the various series expansions for the six cases to the same methods of asymptotic analysis. The series analysed encompass series for the internal energy, the ordering and non-ordering susceptibilities, specific pair correlation functions, and the spherical moments of the pair correlation functions. Assuming the form of the singularities of these functions to be those characteristic of continuous transitions, a variety of different asymptotic analyses have been applied to each series in order to determine critical exponents as well as amplitudes.

A comparative study of the results in the six cases leads to the following observations (Mouritsen 1979): (i) In the cases $\bar{B}_0 \,|\, [001]$ $T > 0$, $T < 0$ and $\bar{B}_0 \,|\, [110]$ $T > 0$, the results are consistent with continuous transitions. The extrapolated series behave smoothly and the physical singularities appear consistently. The critical parameters resulting from different analyses are consistent. (ii) In the cases $\bar{B}_0 \,|\, [111]$ $T > 0$, $T < 0$ (assuming $q = \frac{1}{4}$), the extrapolated series exhibit poor convergence and the results of different analyses are in some cases inconsistent. Expecially the case $\bar{B}_0 \,|\, [111]$ $T < 0$ is irregular and the various Padé approximants often lack physical singularities. (iii) The case $\bar{B}_0 \,|\, [110]$ $T < 0$ (assuming $q = \frac{1}{4}$) is intermediate. The series for the ordering susceptibility and the second spherical moment of the correlation functions are regular, but critical point renormalization indicates that the apparent singularities do no coincide. (iv) Analyses of the critical correlations indicate a break of local symmetry at the transition in the cases $\bar{B}_0 \,|\, [110]$ $T < 0$ and $\bar{B}_0 \,|\, [111]$ $T > 0$. We conclude that series analyses suggest continuous transitions in the double-layer cases and first-order transitions in the multi-layer cases. These conclusions are in accordance with the Monte Carlo results. In the above analyses, we have assumed that the magnetic ground states are of the »permanent« type. For example, in the case $\bar{B}_0 \,|\, [111]$ $T < 0$ this assumption is known to break down. If the series analyses in this case are repeated, now assuming an ordered phase with $q = q^* = 0.2051 \simeq \frac{1}{5}$, the evidence against a continuous transition is removed. The results for the critical exponents, which are included in Table 5.2.1, suggest that the transition is Ising-like.

5.2.10 Nuclear antiferrimagnetic susceptibilities of systems with two spin species: LiF and LiH

The computer study to be reported in this section was carried out in order to answer some open questions in the interpretation of nuclear magnetic resonance experiments on LiF (Cox et al. 1975). The key observable in the resonance experiments is the dispersion signal which can be related to the transverse bulk susceptibility. The susceptibility can be determined for each spin species separately. By comparing the experimental susceptibilities with those calculated from mean-field theory, assuming different types of spin ordering, indirect information regarding the nature of the ordering can be obtained. Obviously, this is only an acceptable procedure in so far as mean-field theory is correct. Cox et al. (1975) encountered a serious discrepancy between the measured and calculated transverse susceptibilities, a discrepancy which was not present for CaF_2. In spite of this discrepancy, it was found that the *ratio* of transverse bulk susceptibilities for the two spin species was in good agreement with mean-field theory. As possible explanations of the deviation between mean-field theory and experiments were given (i) disturbing effects from the paramagnetic impurities, (ii) effects due to sample or field inhomogeneities, and (iii) effects due to the isotope ^6Li with a natural abundance of about 7%. Here we propose (Mouritsen and Knak Jensen 1980a) that, very likely, part of the discrepancy is due to the inadequacy of the mean-field description itself.

The situation we consider is $\overline{B}_o \mid [001]$ at negative temperatures. Only in this case can a polarization be acquired large enough to bring the nuclear spin system into the ordered regime. In this case, the truncated secular dipolar model is a reasonable model producing the correct spin structure. Still, the truncated classical model will have many properties differing from those of the long-range quantum dipolar model. However, we emphasize that the mean-field predictions can still be tested by a comparison with Monte Carlo calculations provided that both types of calculation are performed with reference to the same model. To examine the influence on the results of the ratio, $R = \mu_S/\mu_I$, of magnetic moments, our investigation has been performed for $R = 0.86$ and $R = 0.5$. The former case corresponds to LiH in the classical limit, $\mu_{Li}/\mu_H = 0.86$, and is not much different from the case of LiF, $\mu_{Li}/\mu_F = 0.81$.

The computer simulations are performed on systems with $N = 8^3$ and 14^3 spins. The phase transition is established as continuous for both values of R. Most of the calculations of the bulk susceptibilities are carried out using the smaller lattice. This is acceptable since these non-critical quantities are only slightly influenced by finite-size effects. The transverse bulk susceptibilities are calculated for each spin species individually, $\chi_{x,I}^b$ and $\chi_{x,S}^b$. The total transverse bulk susceptibility is exactly the sum of these two since there is no coupling in the secular dipolar Hamiltonian between the transverse spin components of different spin species. Figure 5.2.8 gives a comparison between the Monte Carlo (MC) and mean-field (MF) results of $\chi_{x,I}^b$ and $\chi_{x,S}^b$ for the two values of R. The susceptibilities are negative since the temperature is negative. It is seen that mean-field theory is qualitatively correct. However, there is a significant quantitative disagreement which increases when R is lowered. For $R = 0.86$, the discrepancy in the ordered phase is for both

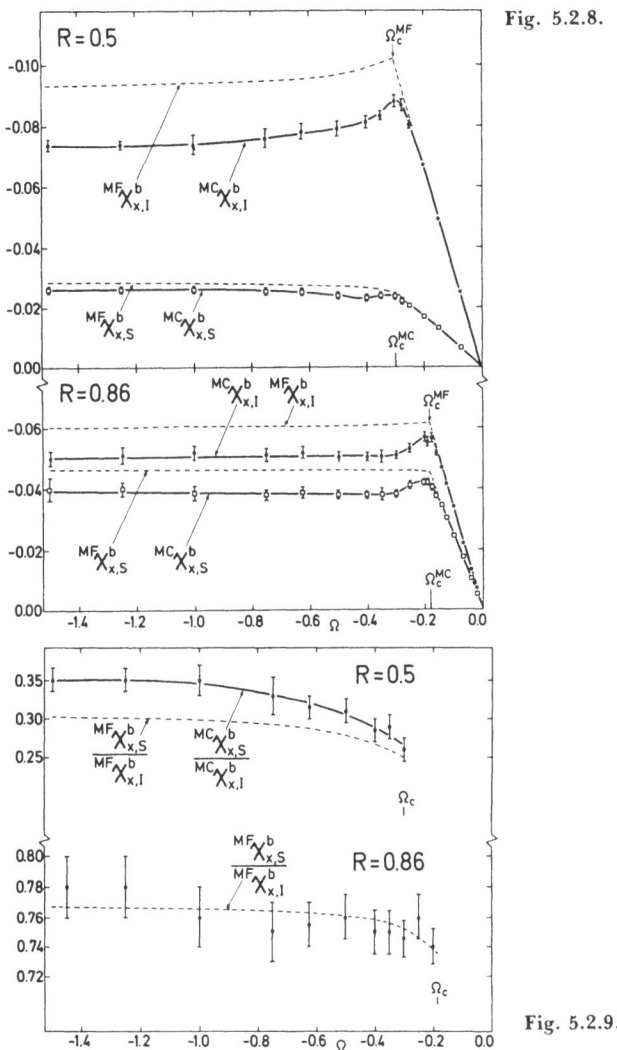

Fig. 5.2.8.

Fig. 5.2.9.

Fig. 5.2.8. Transverse bulk susceptibilities per spin, $\chi^b_{x,I}(T)$ and $\chi^b_{x,S}(T)$, for the two spin species. The figure compares Monte Carlo (MC) and mean-field (MF) calculations. All susceptibilities are in units of r^3_o. $\Omega = \mu^2_I/r^3_o k_B T$.

Fig. 5.2.9. Ratios of transverse bulk susceptibilities in the cases $R = 0.5$ and $R = 0.86$. The figure compares Monte Carlo (MC) and mean-field (MF) calculations, denoted by circles and dashed lines, respectively. $\Omega = \mu^2_I/r^3_o k_B T$.

spin species around 20%. This should be contrasted to a mere 1 – 2% deviation between mean-field and Monte Carlo results for $R = 1$.

From the comparison between mean-field and Monte Carlo results for the transverse bulk susceptibilities shown in Fig. 5.2.8, it can be concluded that the discrepancy between mean-field predictions and the experimental results by Cox et al. (1975) is likely, at least partly, to be due to a shortcoming of the mean-field approach. Cox et al. note that the ratio $^{MF}\chi_{x,S}^b(T)/^{MF}\chi_{x,I}^b(T)$ for LiF agrees with the corresponding experimental ratio within the uncertainty. Figure 5.2.9 shows that such an agreement between mean-field and Monte Carlo results also exists in the entire antiferrimagnetic phase for the truncated secular dipolar model when $R = 0.86$. However, for the smaller value, $R = 0.5$, the mean-field ratio is 20% below the Monte Carlo result.

5.3 Phase Transitions of Adsorbed Monolayers

5.3.1 Two-dimensional phases of molecules adsorbed on solid surfaces

The properties of overlayers of atoms or molecules adsorbed on solid surfaces have within the last five years received an enormous attention among experimentalists as well as theorists. In particular, the structure and the phase transitions of adsorbed monolayers have been of major interest. From a theoretical point of view, the main reason for this interest has been that these layers may be realizations of two-dimensional condensed systems. Theoretically, two-dimensional systems are often more tractable than three-dimensional ones and a large body of theoretical calculations is available for two-dimensional models (see e.g. Sinha (1980) for a review of experimental and theoretical aspects of ordering in two dimensions).

A cleavage surface of a bulk crystalline material presents itself to exterior gas molecules as a regular two-dimensional array of adsorption sites. The degree to which the adsorbate molecules will associate with these sites depends on the competition between the substrate potential and the interactions among the gas molecules. Conventionally, a distinction is made between chemisorption and physisorption depending on whether short-range chemical forces or long-range dispersion forces dominate the substrate potential. Classical examples of the two types of adsorbed systems are oxygen on $Ni(111)$ and noble gases on graphite, respectively. Low-energy electron-diffraction and neutron diffraction studies have shown that the monolayers at appropriate coverages and under suitable thermodynamic conditions may form condensed phases characterized by superstructures which may be in registry or out of registry with the substrate. These phases may undergo phase transitions induced by changes in temperature or coverage. The theories of two-dimensional modulated systems with commensurate, incommensurate, and chaotic phases and their phase transitions have recently been reviewed by Bak (1982). We shall here only be concerned with commensurate phases and order-disorder transitions from these phases to a disordered state.

There is a vast diversity of adsorbed systems undergoing order-disorder transitions. However, the concept of universality imposes a simplifying constraint by hypothesizing that the many different transitions belong to only a small number of universality classes. The most important ones are the universality classes of the ferromagnetic Ising model, the three- and four-state Potts models, the XY

model with cubic anisotropy, and the Heisenberg model with cubic anisotropy (Domany et al. 1978, Domany and Schick 1979, Schick 1982, 1983). Universality then implies that if the Landau-Ginzburg-Wilson Hamiltonian (cf. Sec. 4.1.1) of the physical system under consideration is the same as that of one of the models listed above, the system has a phase transition of the same nature as that model. Specifically, if the transition is continuous, the critical exponents will be the same.

Statistical surface physics and the order-disorder phase transitions of commensurate adsorbed monolayers are conveniently studied by means of lattice models.[*] The lattice sites represent the preferred adsorption sites. Since we are not interested in properties of the substrate itself, this may simply be thought of as a symmetry-breaking field which, in combination with the interactions within the monolayer, gives rise to the lattice and an effective anisotropic interaction between the adsorbate molecules.

Computer simulation has proved to be an important theoretical technique to unravel the phase behavior of models for adsorbed monolayers. Simulations have been used both to fit simple models to diffraction intensities (Williams et al. 1978, Roelofs et al. 1979, Ching et al. 1978; for further references, see Weinberg 1983), to investigate the phase transitions of realistic models of specific systems (Abraham 1982, Secs. 5.3.4 and 5.3.6), and to study two-dimensional pure models as simple representatives of important universality classes (Binder and Landau 1980, Glosli and Plischke 1983; see also Landau 1979b for a list of references). A Monte Carlo study of a realistic microscopic interaction model of N_2 adsorbed on graphite will be described in Secs. 5.3.4 and 5.3.6. Computer simulation studies of the very process of forming two-dimensional ordered structures and the kinetics of domain growth in two dimensions is the subject of Sec. 5.4.

5.3.2 N_2 physisorbed on graphite: The anisotropic-planar rotor model

Diatomic molecules are interesting adsorbates since they have translational as well as internal (orientational) degrees of freedom. This additional feature gives rise to a remarkably rich phase diagram including phases with doubling of the unit cell. Here, we shall be concerned with commensurate phases only. The class of adsorbates we want to discuss includes N_2, H_2, and D_2. Physisorbed on the basal plane of graphite, these gases form $\sqrt{3} \times \sqrt{3}$ monolayer condensed phases (Nielsen et al. 1977, Kjems et al. 1976). In the following, we concentrate on N_2 and offer a few remarks on H_2 and D_2 towards the end.

At temperatures below 47K and at coverages less than $\frac{1}{3}$ of the number of hexagonal sites of the substrate, N_2 molecules form a registered $\sqrt{3} \times \sqrt{3}$ commensurate overlayer structure on graphite. The centers of mass of the molecules constitute a triangular lattice with the lattice points being at the centers of the honeycombs of the substrate mesh (Kjems et al. 1976). Thus, the translational degrees of freedom are effectively discretized and the problem is separated from two-dimensional melting. The triangular lattice of absorption sites has a lattice

---

[*] Incommensurate-commensurate transitions may also be studied via lattice models, see e.g. Selke et al. 1983, Saito 1981.

spacing of $r_o = 4.26\text{Å}$, i.e. $\sqrt{3}$ times larger than that of the graphite mesh. Obviously, the $\sqrt{3} \times \sqrt{3}$ structure may be placed on three equivalent positions on the honeycomb substrate. As the temperature is decreased below 30K, a further transition occurs which was first detected by specific heat measurements (Chung and Dash 1977). Later, neutron scattering studies (Eckert et al. 1979) showed that this transition involves the formation of a superlattice structure with a doubling of the unit cell along at least one of the crystal axes. Quite recently, low-energy electron-diffraction measurements (Diehl et al. 1982) have led to the unambiguous identification of the low-temperature phase as a 2×1 *herringbone* structure as shown schematically in Fig. 5.3.1. The transition around 30K thus proceeds via the orientational degrees of freedom.

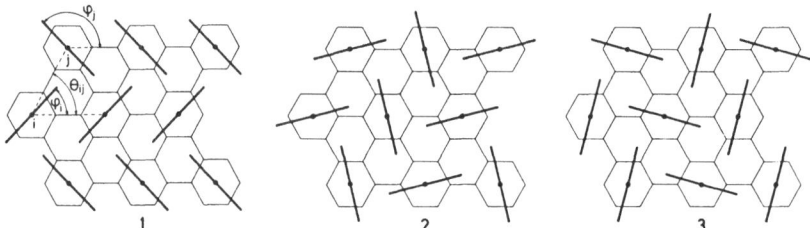

Fig. 5.3.1. Schematic representation of the three degenerate herringbone domains on a triangular lattice. The planar rotors represent the interatomic axes of diatomic molecules (e.g. N_2) physisorbed in a commensurate $\sqrt{3} \times \sqrt{3}$ overlayer on the hexagonal substrate of graphite. The angles φ_i and θ_{ij} enter the Hamiltonian of the anisotropic-planar rotor model in Eq. (5.3.1).

The herringbone structure and its thermodynamic properties are determined by the anisotropic interactions of the N_2 molecules with the substrate and with each other. In the case of N_2, the substrate potential is very large and negative (Steele 1977). Therefore, the interatomic axes are effectively confined to the graphite plane. The intermolecular interactions are, as in solid N_2, well described by classical electric quadrupole-quadrupole interactions (Scott 1976). That these interactions indeed have the herringbone structure as the ground state has been demonstrated by Monte Carlo calculations (O'Shea and Klein 1979) and by energy minimization calculations based on the electric quadrupole-quadrupole interaction and atom-atom potentials (Fuselier et al. 1978). For planar quadrupoles on a triangular lattice, this interaction takes the form of an anisotropic-planar rotor Hamiltonian (Mouritsen and Berlinsky 1982)

$$H = J \sum_{<i,j>} \cos(2\varphi_i - 2\varphi_j) + K \sum_{<i,j>} \cos(2\varphi_i + 2\varphi_j - 4\theta_{ij}), \qquad (5.3.1)$$

where the summations are over the six nearest-neighbor pairs only. φ_i measures the orientation of the ith molecule and θ_{ij} specifies the direction of the line connecting sites i and j of the lattice (cf. Fig. 5.3.1). For pure electric quadrupole-quadrupole interactions, the ratio of the coupling constants is $K/J = 35/3$ and the isotropic XY-like term in Eq. (5.3.1) can therefore be neglected. It is the remaining anisotropic term which stabilizes the herringbone structure. The coupling constant K is related to the nearest-neighbor electric quadrupole-quadrupole coupling constant Γ_0 through the relation

$$K = \frac{875}{64}\Gamma_0 = \frac{875}{64}\frac{6(eQ)^2}{25r_0^5} \simeq (33 \pm 5)\text{K}, \tag{5.3.2}$$

where we have used that the electric quadrupole moment of N_2 is $eQ = (1.4 \pm 0.1) \times 10^{-26}$esu cm^2 (Buckingham et al. 1968). With the value of K given in Eq. (5.3.2) and with $J = 0$, the anisotropic-planar rotor Hamiltonian is believed to be a highly realistic lattice model (with *no* adjustable parameters) of the orientational properties of N_2 physisorbed on graphite.

The order parameter of the 2×1 herringbone structure has *three* real components corresponding the the three-fold symmetry axis of the triangular lattice, cf. Fig. 5.3.1. Thus, there are six thermodynamically equivalent ground states. The components of the order parameter may be written as

$$\psi_\alpha(T) = N^{-1} < | \sum_{i=1}^{N} \sin(2\varphi_i - 2\theta_\alpha)e^{i\bar{q}_\alpha \cdot \bar{r}_i} | >, \quad \alpha = 1, 2, 3, \tag{5.3.3}$$

where $\bar{q}_\alpha$ are the propagation vectors of the herringbone structure and θ_α are the directional angles of the triangular lattice

$$\bar{q}_1 = (0, \frac{2\pi}{r_0\sqrt{3}}), \quad \theta_1 = 0$$

$$\bar{q}_2 = (-\frac{\pi}{r_0}, -\frac{\pi}{r_0\sqrt{3}}), \quad \theta_2 = \frac{2\pi}{3} \tag{5.3.4}$$

$$\bar{q}_3 = (\frac{\pi}{r_0}, -\frac{\pi}{r_0\sqrt{3}}), \quad \theta_3 = \frac{4\pi}{3}.$$

The nature of the phase transition from the herringbone phase to the orientationally disordered phase will be discussed in Sec. 5.3.4 below.

The anisotropic-planar rotor model in Eq. (5.3.1) represents the limit of infinite negative substrate potential, V_c. The complete phase diagram, spanned by V_c and the temperature, has been calculated within the mean-field theory by Harris and Berlinsky (1979) in the case of the full electric quadrupole-quadrupole interaction. Part of the phase diagram is shown in Fig. 5.3.2. According to this diagram, the out-of-plane component of the quadrupoles comes to play a role as V_c is increased and eventually a new phase, the *pinwheel* phase, possessing a different symmetry becomes stable. The pinwheel structure, which is shown schematically in Fig. 5.3.3, is a linear superposition of equal amplitudes of each of the three herringbone domains of Fig. 5.3.1. The pinwheel phase has *four* order parameter components and thus eight degenerate ground states. Such a phase is usually denoted a $p(2 \times 2)$ phase.

Experimentally, it is difficult to explore the phase diagram spanned by V_c and T since, in a real system, V_c cannot be adjusted at will. However, Harris et al. (1984)

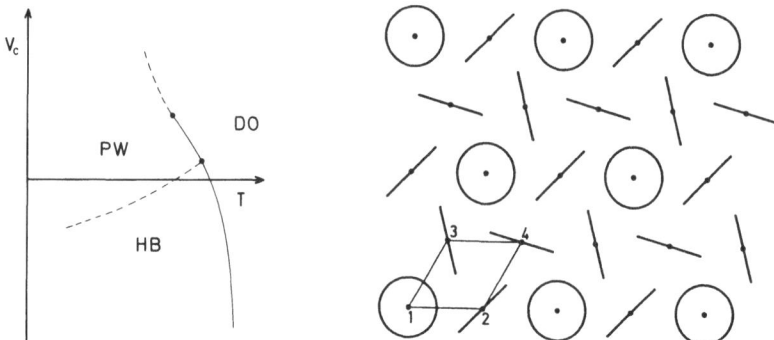

Fig. 5.3.2. Part of the mean-field phase diagram for quadrupoles on a triangular lattice. V_c is the substrate potential. HB, PW, and DO refer to herringbone, pinwheel, and disordered phases. First-order and continuous transitions are indicated by dashed and solid lines, respectively.

Fig. 5.3.3. Schematic representation of the pinwheel structure of quadrupoles on a triangular lattice. The unit cell and the four sublattices of the ordering are indicated. In the case of the full electric quadrupole-quadrupole interaction, the circles indicate quadrupoles which are disordered. In the case of the annealed dilution of planar quadrupoles coupled by Eq. (5.3.5), the circles indicate vacancies.

have recently pointed out that it is possible to escape this difficulty by studying an *annealed dilution* of planar quadrupoles on a triangular lattice coupled by the modified anisotropic-planar rotor Hamiltonian

$$H = K \sum_{\langle i,j \rangle} x_i x_j \cos(2\varphi_i + 2\varphi_j - 4\theta_{ij}). \qquad (5.3.5)$$

In Eq. (5.3.5), $x_i = 0, 1$ is the single-site occupation variable subject to the constraint

$$\sum_{i=1}^{N} x_i = xN, \qquad (5.3.6)$$

where x is the concentration (coverage) of the rotors. For the annealed dilution, the chemical potential of the vacancies turns out to play the role of V_c in Fig. 5.3.2 (Harris et al. 1984). A possible physical realization of this situation is a mixture of N_2 and rotationally inert rare gases (Kr or Ar) on graphite.[*] Determination of the phase diagram of the anisotropic-planar rotor Hamiltonian on a triangular lattice with vacancies is the subject of Sec. 5.3.6.

5.3.3 The Heisenberg model with cubic anisotropy

To facilitate the discussion of the phase transitions of the anisotropic-planar rotor model, it is useful to make a brief discursion to introduce the Heisenberg model with cubic anisotropy. By this model, two universality classes pertinent to surface

---

[*] S. Satija and L. Passell, preliminary experiments.

phase transitions may be defined. The Heisenberg model with cubic anisotropy is characterized by the Landau-Ginzburg-Wilson Hamiltonian (see e.g. Schick 1983)

$$H(\overline{\psi}) = r \sum_{\alpha=1}^{3} \psi_\alpha^2 + \sum_{\alpha=1}^{3} (\nabla \psi_\alpha)^2 + u(\sum_{\alpha=1}^{3} (\psi_\alpha)^2)^2 + v \sum_{\alpha=1}^{3} (\psi_\alpha)^4 \qquad (5.3.7)$$

with $\alpha = 1, 2, 3$ refering to the three (Heisenberg) components of the ordering field $\overline{\psi} = (\psi_1, \psi_2, \psi_3)$. The last term in Eq. (5.3.7), which is invariant only under the symmetry operations of the cube, breaks the rotational symmetry of $\overline{\psi}$. Under a renormalization group transformation, this term will become dominant implying that only the stationary points of the cube will play a role (Domany and Riedel 1979, Nienhuis et al. 1983, Pfeuty and Toulouse 1977). Thus, $\overline{\psi}$ will point either to the faces or the corners of the cube. In these two cases, the fourth-order term will take on the values $(u + v)|\overline{\psi}|^4$ and $(u + \frac{v}{3})|\overline{\psi}|^4$. Both these fourth-order coefficients have to be positive in order for the Hamiltonian to stay thermodynamically stable under the renormalization group transformation. For $v < 0$, the faces are favored and for $v > 0$, the corners are favored. This leads to the notion of face- and corner-type cubic anisotropy.

The two types of cubic anisotropy give rise to two universality classes. Renormalization group calculations predict that in the case of face-type cubic anisotropy, the phase transition is always of first order (Pfeuty and Toulouse 1977, Aharony 1973a, Ketley and Wallce 1973). In the case of corner-type anisotropy, the transition may either be continuous or of first order. If it is continuous, the critical behavior is predicted to be that of the two-dimensional Ising model (Schick 1983, Grest and Widom 1981).

It can be shown that the Landau-Ginzburg-Wilson Hamiltonian, which may be constructed from the herringbone order parameters in Eq. (5.3.3), is that of the Heisenberg model with face-type cubic anisotropy (Harris and Berlinsky 1979). Similarly, the order parameters of the pinwheel structure lead to a Landau-Ginzburg-Wilson Hamiltonian which has corner-type cubic anisotropy. By universality, we are therefore faced with the theoretical prediction of a first-order transition for the herringbone phase and a possible continuous transition (with Ising exponents) of the pinwheel phase. We now proceed to test these predictions.

5.3.4 Fluctuation-induced first-order phase transition in the anisotropic-planar rotor model

In this section, we report the results of a computer simulation study (Mouritsen and Berlinsky 1982) of the phase transition in the anisotropic-planar rotor model in Eq. (5.3.1). This study serves several purposes. First of all, it supplies numerical experimental information on a realistic model of an adsorbed system which is extremely difficult to study in detail by conventional experiments. Secondly, it provides a test of the renormalization group prediction of first-order transitions in the universality class of the two-dimensional Heisenberg model with face-type cubic anisotropy. Finally, it allows for an interpretation of special effects observed in low-energy electron-diffraction studies of N_2 on graphite.

Compared to real experiments, computer simulations have a number of advantages (cf. Sec. 2.2.11) which are particularly favorable when studying adsorbed monolayer systems. Most important is it that the simulations are carried out on truly two-dimensional systems with no three-dimensional character introduced by a substrate. Moreover, the ordered structures produced in the simulations are not influenced by disturbing impurities, desorption effects, or inhomogeneities of the substrate. The drawback of finite-size effects in computer simulations may be turned into an advantage when a comparison with experimental data is performed. Often, the coherence of the substrate only extends to sizes comparable with the largest lattices used in current computer work.

The main purpose of the present section will be to discuss the thermodynamics of the anisotropic-planar rotor model itself. A comparison with experimental measurements will be made in Sec. 5.3.5. Mouritsen and Berlinsky (1982) used conventional Monte Carlo techniques to calculate, for a series of different lattice sizes with periodic boundary conditions, the internal energy and the order parameter components, Eq. (5.3.3), of the anisotropic-planar rotor model. A Glauber-type excitation mechanism is used corresponding to rotation of the individual rotor through an arbitrary angle, $\varphi_i \to \varphi_i + \Delta\varphi_i$, $0 \leq \Delta\varphi_i \leq \pi$. The thermal equilibrium values are evaluated by using coarse-graining techniques as well as distribution functions (cf. Sec. 2.2.6). In Fig. 5.3.4 are shown the results for the temperature dependence of the internal energy. The energy is seen to change dramatically in a narrow temperature region. Hardly any finite-size effects are observed. The specific heat, $C(T)$, calculated from the fluctuation theorem, Eq. (2.1.7), is shown in Fig. 5.3.5 for two small lattices with $N = 400$ and 1600 sites. $C(T)$ has a

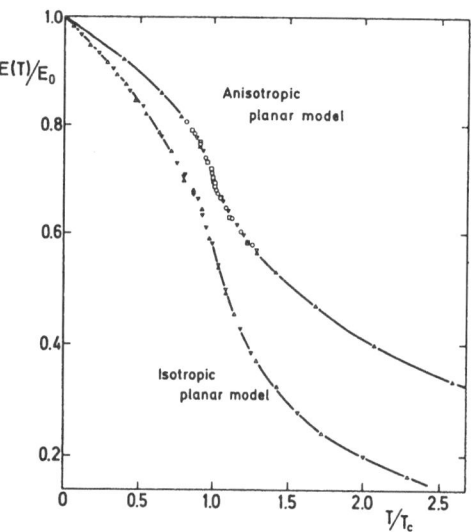

Fig. 5.3.4. Normalized internal energy for the anisotropic-planar rotor model and for the isotropic-planar rotor model as defined by Eqs. (5.3.1) and (5.3.8). The data are obtained from Monte Carlo calculations on lattices with N sites. $N = 400$ (triangles), 1600 (inverse triangles), 6400 (circles), 10000 (squares). The temperature is scaled with the transition temperature, T_c, obtained from the position of the specific heat maximum.

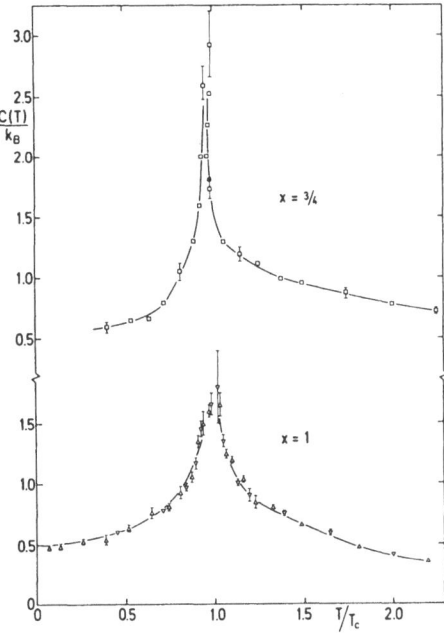

Fig. 5.3.5. Specific heat, $C(T)$, per planar rotor of the anisotropic-planar rotor model for coverages $x = \frac{3}{4}$ and $x = 1$. The temperatures are scaled with the appropriate transition temperature, T_c. The data points refer to Monte Carlo calculations on lattices with $N = 400$ (inverse triangles), $N = 1600$ (triangles), and $N = 10000$ (squares) sites.

pronounced and broadened peak at the temperature where the internal energy has its inflection point. It is observed that $C(T)$ is insensitive to N for $N \gtrsim 400$, both with respect to position and height of the peak. Thus, the results represent the thermodynamic limit. In accordance with the equipartition theorem, the classical property of the model makes $C(T)$ approach $k_B/2$ at low temperatures.

The data presented so far form no conclusive evidence of a conventional finite-temperature order-disorder transition in the anisotropic-planar rotor model. In fact, the data look very similar to those obtained for the *isotropic*-planar rotor model (or XY model) which is defined by

$$H = -J \sum_{<i,j>} \cos(\varphi_i - \varphi_j). \qquad (5.3.8)$$

The isotropic model is predicted not to support conventional long-range order in two dimensions (Villain 1975, José et al. 1977, Kosterlitz and Thouless 1973), but rather to have a low-temperature phase characterized by a power-law decay of the pair correlation function modified by the presence of pairs of topological defects that are spin vortices of opposite polarity. An extensive Monte Carlo study of the isotropic-planar rotor model on the square lattice has confirmed these predictions (Tobochnik and Chester 1979; see also Kawabata and Binder 1977, Shugard et al. 1980). The temperature dependence of the internal energy[*] for the isotropic

---

[*] O.G. Mouritsen (unpublished calculations)

model on a triangular lattice is also shown in Fig. 5.3.4. The tangent in the inflection point is less steep than for the anisotropic model. More important is it that the specific heat, of the isotropic-planar rotor model on both the triangular and square lattices exhibits a broadened peak quantitatively similar to that of the anisotropic model. Also for the isotropic model, $C(T)$ appears to be insensitive to the lattice size. It should be noted that the specific heat anomaly of the isotropic-planar rotor model does not signal a phase transition. In the case of the square lattice, this peak anomaly occurs about 15% above the vortex-unbinding Kosterlitz-Thouless transition (Tobochnik and Chester 1979).

The conclusive numerical evidence of a conventional order-disorder transition in the anisotropic-planar rotor model comes from a study of the thermal decay of the order parameter $\psi(T)$. For $k_B T/K \lesssim 0.775 \equiv k_B T_c/K$, the system is found to be ordered in one of the three components, ψ_α Eq. (5.3.3), which is then termed *the* order parameter. The two remaining components are equal, but finite, because of finite-size effects. The Monte Carlo data for $\psi(T)$ are shown in Fig. 5.3.6. For $T \gtrsim T_c$, the finite-size order also resides in one of the components close to T_c, but not far above T_c the order is equally distributed among the three components. For $T \lesssim 0.95 T_c$, there is no observable finite-size effects on $\psi(T)$. This is in marked contrast to the Monte Carlo results[*] of the »order parameter« for the isotropic

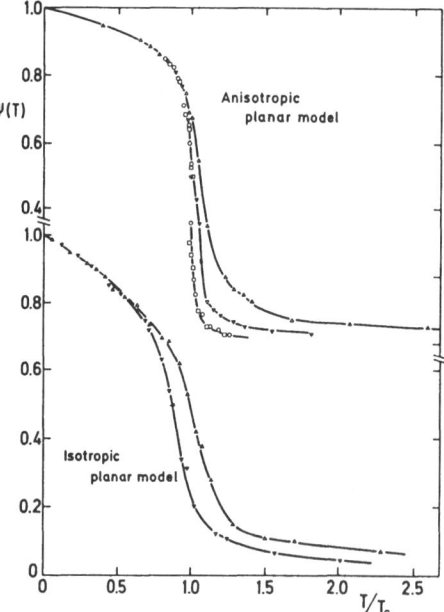

Fig. 5.3.6. Temperature dependence of the order parameter, $\psi(T)$ Eq. (5.3.3) and (5.3.9) for the anisotropic-planar rotor model and for the isotropic-planar rotor model defined in Eqs. (5.3.1) and (5.3.8). The results are obtained from Monte Carlo calculations on systems with N rotors. $N = 400$ (triangles), 1600 (inverse triangles), 6400 (circles), and 10000 (squares). The temperature is scaled with the appropriate transition temperature, T_c, obtained from the position of the specific heat maximum.

[*] O.G. Mouritsen, unpublished calculations.

model

$$\psi(T) = N^{-1} < [(\sum_{i=1}^{N} \cos\varphi_i)^2 + (\sum_{i=1}^{N} \sin\varphi_i)^2]^{\frac{1}{2}} > \qquad (5.3.9)$$

which are also shown in Fig. 5.3.6. $\psi(T)$ shows a clear finite-size dependence down to low temperatures. A thorough finite-size analysis of $\psi(T)$ for the isotropic model on the square lattice (Tobochnik and Chester 1979) demonstrates that at low temperatures the long-range order is lost as $\psi^2(T) \sim N^{-T/4\pi}$ in agreement with spin-wave theory. This result is not expected to depend on lattice structure. From Fig. 5.3.6, we may thus conclude that the anisotropic planar rotor model has a finite-temperature phase transition of conventional order-disorder type. The transition temperature, T_c, estimated from the order parameter behavior corresponds to the temperature at which the specific heat attains its maximum.

The nature of this phase transition is revealed by studying the size-dependence in the transition region. For $N = 400$ and 1600, $\psi(T)$ varies continuously through T_c. Conversely, for $N = 6400$ there is a slight indication of a discontinuity at T_c, although the jump is somewhat smeared because of a rapid shifting of the order between the three components. However, for $N = 10000$ a clear discontinuity has developed. In Fig. 5.3.7, an enlargement of the critical region is given and the discontinuity in $\psi(T)$ is now clearly exposed. Furthermore, this figure shows that the transition is associated with hysteresis of width $\Delta T/T_c \simeq 1.5\%$. The hysteresis is completely smeared for the smaller lattices. The discontinuity at T_c increases with N. The way in which the order parameter in the transition region may be obtained from the distribution functions of the individual order parameter components is illustrated in Fig. 5.3.8 (cf. Sec. 2.2.6). This figure clearly shows that the value of the order parameter at T_c depends on the thermal history of the

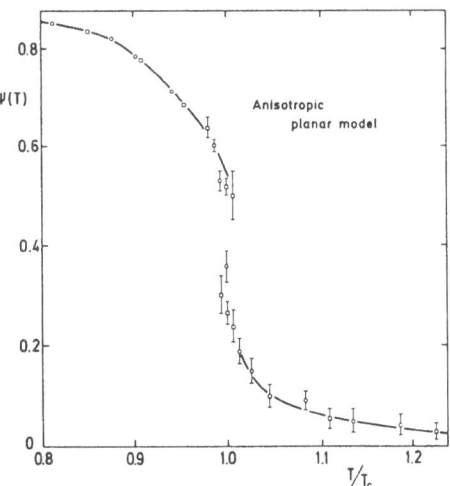

Fig. 5.3.7. Temperature dependence of the order parameter, Eq. (5.3.3), of the anisotropic-planar rotor model in the transition region. The temperature is scaled with the transition temperature, T_c. The results are obtained from Monte Carlo calculations on systems with $N = 6400$ (circles) and $N = 10000$ (squares) rotors.

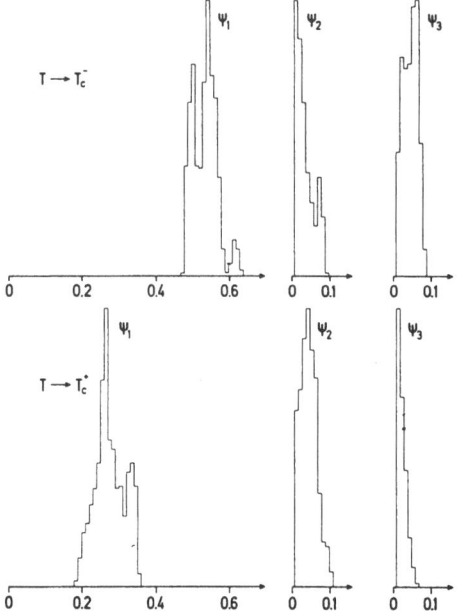

Fig. 5.3.8. Distribution functions of the three order parameter components, $\psi_\alpha(T)$ Eq. (5.3.3), of the anisotropic-planar rotor model as the transition temperature, T_c, is approached from above and below. The data are derived from Monte Carlo calculations on a lattice with $N = 10000$ rotors.

system. All this evidence suggests that the transition from the herringbone phase to the orientationally disordered phase is of first order. The pronounced finite-size tails on $\psi(T)$ above the transition (cf. Fig. 5.3.6) as well as the fact that very large systems are required to resolve the discontinuity at T_c indicate that this first-order transition is strongly influenced by fluctuations.

The finding of a first-order phase transition lends support to the renormalization group prediction that phase transitions belonging to the universality class of the two-dimensional Heisenberg model ($n = 3$) with face-type cubic anisotropy are of first order (Sec. 5.3.3). Since the mean-field theory predicts a continuous transition for the anisotropic-planar rotor model (cf. Fig. 5.3.2 for $V_c \to -\infty$), we furthermore arrive at the conclusion that the observed first-order phase transition is *fluctuation induced* (cf. Sec. 4.1.1).[*] Schick (1983) has argued that the fluctuations may be caused by the presence of a marginal operator at $n = 2$. The situation thus parallels that of $q = 5$- and 6-state two-dimensional Potts models where the strong fluctuations found near the first-order transitions (Binder 1981b) may be caused by the marginal operator at $q = 4$.

---

[*] Recently, Glosli and Plischke (1983) have reported on a Monte Carlo study of the Ising model with nearest and next-nearest neighbor interactions on a triangular lattice. The transition from the 2×1 phase to the paramagnetic phase of this model also belongs to the universality class of the Heisenberg model with face-type cubic anisotropy. In agreement with our results, these authors also find a first-order transition for this symmetry.

Two other computer simulation studies have been reported on the orientational phases of classical quadrupoles on a triangular lattice in the presence of a large negative substrate potential. The first one is that of O'Shea and Klein (1979) who used the full interaction Hamiltonian of planar quadrupoles. In this study, only very small lattices ($N = 6 \times 6$ and 12×12) were investigated and it is therefore impossible to draw any conclusions regarding the nature of the orientational transition. However, the behavior of the internal energy and the specific heat indicates the presence of a transition around $k_B T/K \simeq 0.83$. This value is in reasonable agreement with our estimate for the anisotropic-planar rotor model, considering that we have left out the isotropic part of the quadrupole-quadrupole interaction in Eq. (5.3.1).

The second computer simulation study (Migone et al. 1983) is based on a molecular dynamics calculation for a three-dimensional assembly of 96 N_2 molecules confined in a rectangular cell. The density corresponds to the coverage $x = 1$. The molecules have translational degrees of freedom and the full expression is used for the substrate potential and the $N_2 - N_2$ quadrupolar interactions (Steele 1977). Thus, this model reproduces more faithfully the true interactions of the adsorbed system than does the anisotropic-planar rotor model. The advantages of using such a realistic model are obvious. However, the complexity of the Hamiltonian precludes simulations on large systems and the molecular dynamics study therefore excludes itself from obtaining reliable information in the transition region. Nevertheless, the specific heat obtained from the molecular dynamics calculation, although of low accuracy, displays a pronounced peak indicative of a phase transition. Migone et al. also calculated various orientational order parameters of which we shall focus on the one, $\bar{\psi}(T)$, corresponding to the herringbone order parameters in Eq. (5.3.3). In their definition of $\bar{\psi}(T)$, however, these authors did not use the absolute value of the instantaneous order parameter in calculating the ensemble average. Thus, according to the discussion in Sec. 2.2.5, $\bar{\psi}(T)$ is only a proper measure of the long-range order as long as the ergodicity of the system is effectively broken. The smaller the system, the lower the temperatures which are required to fulfil this criterion (refering to a common observation time). For a system with only 96 particles, an ergodic behavior in computer simulations is expected down to temperatures considerably below the phase transition. Therefore, we consider the findings by Migone et al. of a rapidly vanishing long-range order as T_c is approached from below and of a lack of finite-size rounding above T_c as clear indications of an ill-defined long-range order parameter. In spite of this deficiency in the molecular dynamics study, there are two important new results which derive from these calculations. The first one is the finding that the substrate potential induces a hexagonal ordering of the quadrupoles at low temperatures and that there is no sharp loss in translational order through the transition. This observation corroborates the use of lattice models for the transition and demonstrates convincingly that the transition proceeds via orientational degrees of freedom only. The second new result is that the out-of-plane fluctuations become important as the temperature is raised, but no sharp increase is observed at the transition.

Before we turn to a comparison between the results of model calculations and experimental measurements of N_2 adsorbed on graphite, brief reference should be given to three related computer simulation studies of orientational ordering transitions. The first one is a Monte Carlo study of a bilayer lattice model of classical quadrupoles (O'Shea and Klein 1982) modelling N_2 on graphite above unity coverage where a second layer is built up (Diehl and Fain 1983). In agreement with experiments, the simulations show that the lower layer maintains its 2×1 herringbone ordering at low temperatures. The second study is Cotterill's (1976) molecular dynamics study of the two-dimensional dumbbell model of head group properties of a lipid bilayer (cf. Sec. 5.1.6). This model has translational degrees of freedom and the dumbbells interact via Lennard-Jones potentials as well as via a Coulomb interaction between identical charges assigned to each dumbbell. A very interesting melting transition is observed which is mediated by orientational disclination lines. The model reproduces the experimentally observed dependence on pH of the transition temperature of charged lipid membranes. The simulations are not extensive enough to provide details about the phase transition. The third study is a molecular dynamics calculation on a square lattice model of diatomic molecules interacting via Lennard-Jones potentials (Kalia et al. 1982). The model corresponds to diatomic molecules adsorbed on substrates such as MgO and LiF. No experimental information is available as yet on such systems. The model has a one-component orientational order parameter. For lattices with up to 900 sites, no indications of a first-order transition is found. Kalia et al. conclude that the transition is continuous. A few remarks are in order on these results. Firstly, the computer simulations on the anisotropic-planar rotor model show (cf. Fig. 5.3.6) that a transition which looks continuous in small lattices may eventually in larger lattices turn out to be of first order. Kalia et al. point to the observed exponential relaxation behavior as characteristic of a continuous transition. However, the relaxation behavior of a small system of anisotropic-planar quadrupoles has a similar exponential behavior due to the large critical-like fluctuations. The second remark to make is that by symmetry the transition in the Lennard-Jones system belongs to the Ising universality class and thus a continuous transition is expected.

5.3.5 Comparison with experiments on N_2 physisorbed on graphite

The first microscopic experimental investigation by neutron scattering (Eckert et al. 1979) of the phase transition around 30K in monolayers of N_2 on graphite provided neither any unambiguous identification of the ordered phase nor any details about the nature of the phase transition. It was first with the low-energy electron-diffraction studies performed by Diehl et al. (1982) that the low-temperature herringbone structure could be identified. The key observation leading to this identification is that not only does the diffraction pattern show superlattice peaks (as in the neutron study) indicating a doubling of the unit cell, but the absence of certain reflections demonstrates that the structure has two perpendicular glide lines. These facts point unambiguously towards a 2×1 herringbone structure (for a review of the experiments, see Diehl and Fain 1983). The theoretical model predictions derived from computer simulations (O'Shea and Klein 1979, Mouritsen and Berlinsky 1982) and from energy minimization calculations (Fuselier et al. 1978) of a low-temperature herringbone phase are thus in accordance with experiments.

As for the nature of the phase transition, the low-energy electron-diffraction experiments cannot completely exclude a continuous transition. The diffraction intensities change dramatically, but in a continuous manner, through the transition, and the superlattice intensity persists up to about 40K. The intensity above 30K changes from run to run on the same sample and increases as a function of time. Diehl et al. (1982) suggest that these observations may indicate a first-order transition masked by finite-size effects which are being induced by a combination of substrate inhomogeneity and »long-ranged« short-range order above the transition. For this interpretation, the Monte Carlo data in Fig. 5.3.6 of the herringbone order parameter as a function of temperature and system size play an important role (Diehl and Fain 1983). The graphite surface may have regions which are coherent over thousands of Å. Still, the resolution of the low-energy electron-diffraction experiments corresponds to a coherence length of about 100Å. The linear dimensions of the systems studied in the Monte Carlo simulations (Mouritsen and Berlinsky 1982) range from 80Å to 400Å. Even for the largest system simulated, a pronounced high-temperature tail is observed (cf. Fig. 5.3.6). Since the diffraction experiments can only determine that the ordered domains above T_c are not smaller than 80Å, a consistent interpretation of the sharp intensities above T_c would be to associate them with scattering from a distribution of small domains which are orientationally ordered due to finite-size effects. The breaking up into domains may in turn be caused by substrate inhomogeneities.

Direct experimental evidence in favor of a first-order transition comes from a recent ^{15}N-nuclear magnetic resonance study of monolayers of $^{15}N_2$ on graphite (Sullivan and Vaissiere 1983). ^{15}N rather than ^{14}N was studied in order to avoid broadening due to nuclear quadrupolar splitting. Being a purely local technique, the resonance experiment cannot directly establish the nature of the ordered state. On the other hand, local order parameters may reflect the phase transition in the same way as a pair correlation function can. Indeed, the resonance experiment shows a distinct discontinuity of the local order parameter at a temperature $T_c = (28.1 \pm 0.1)$K thus signalling a first-order phase transition. The transition is associated with hysteresis extending about 2.5% around T_c. These observations are in excellent agreement with the Monte Carlo results of Sec. 5.3.4 considering that the precise width of the hysteresis loop depends on the observation time.

Migone et al. (1983) have attempted to determine the nature of the herringbone transition from high-precision specific heat data. By analyzing the specific heat peak in terms of a singularity $|T - T_c|^{-\alpha}$, these authors arrive at a value of the exponent $\alpha \simeq 1.3 \pm 0.2$ inconsistent with critical-point behavior. Since neither a background constant has been allowed for in this analysis nor are the data reliable below $|T - T_c| \simeq 10^{-2}$, we do not consider this analysis as providing convincing evidence of a first-order transition. The high-precision calorimetric measurements of the specific heat (Migone et al. 1983) reveal the specific heat peak as highly symmetric. This is in qualitative agreement with the Monte Carlo results for the anisotropic-planar rotor model in Fig. 5.3.5. However, the heat content in the experimental peak is larger than that in the simulated peak. This is probably due to a too severe restriction of the quadrupolar rotations implied by the planar rotor model in Eq. (5.3.1) rather than to the suppression of translational degrees of freedom. Support for this suggestion is provided by the molecular dynamics calculation on the non-lattice model allowing for out-of-plane rotations, which leads to a heat content of the specific heat almost twice of that found experimentally (Migone et al. 1983).

The experimental values of the transition temperature, T_c, of the herringbone transition is found in the range from 28K to 30K depending on the experimental technique. The value of $T_c = (28.1 \pm 0.1)$K obtained from magnetic resonance experiments appears to be the most reliable estimate (Sullivan and Vaissiere 1983). The Monte Carlo result for T_c of the anisotropic-planar rotor model is found to be $T_c \simeq 26$K by using the value of the quadrupole-quadrupole coupling constant in Eq. (5.3.2). O'Shea and Klein's computer study leads to $T_c \simeq 28$K and the molecular dynamics calculation of Migone et al. yields $T_c \simeq 34$K. Thus it appears that the various microscopic models used to model the herringbone transition produce fairly reasonable transition temperatures. By comparing the results of T_c for the anisotropic-planar rotor model and for the planar rotor model including the isotropic term (O'Shea and Klein 1979), we conclude that neglect of the isotropic term in Eq. (5.3.1) leads to a suppresion of T_c by a few degrees. The uncertainty in the experimental value of the quadrupole moment of N_2, eQ, is too large to determine which one of the models are the more realistic with respect to a determination of the phase transition temperature.

5.3.6 Phase behavior of the anisotropic-planar rotor model with vacancies

In a combined theoretical study including Monte Carlo calculations, mean-field theory, and spin-wave theory, Harris et al. (1984) have analyzed the phase diagram of a system of planar rotors on a triangular lattice with annealed vacancies. The rotors interact via the anisotropic-planar rotor model in Eq. (5.3.5). In the present section, we shall describe the phase behavior of this model in terms of a phase diagram spanned by temperature T and coverage x. In Sec. 5.3.7, a number of physical realizations of the model will be discussed.

That the pinwheel structure of Fig. 5.3.3 gains stability when vacancies are introduced into the anisotropic-planar rotor model may be appreciated when noting that the herringbone structure is frustrated in the sense that not all bond energies are saturated. Bonds within a sublattice of the herringbone structure are at minumum energy, $-K$, whereas bonds between different sublattices have energy $-\frac{K}{2}$. The ground state energy per molecule is thus $-2K$ which is exactly the same as that of the pinwheel structure which has all bonds saturated at the minumum energy $-K$. The chemical potentials of the two phases are therefore equal at zero temperature. The relative stability of the two phases at finite temperatures represents a more complicated problem which we shall address next.

The Monte Carlo calculations have been performed along a single path of constant coverage, $x = \frac{3}{4}$, in the phase diagram. As pointed out in Sec. 5.3.3, the phase behavior at this coverage is of special interest because, if a single phase line separates the pinwheel phase from the orientationally disordered phase, then the phase transition belongs to the universality class of the two-dimensional Heisenberg model with corner-type cubic anisotropy. The computer simulations are carried out on a series of rectangular-shaped finite triangular lattices with periodic boundary conditions. The lattices have $N = L^2$ sites, with $L = 20, 28, 40,$ and 100. The system of interacting rotors is brought to the *annealed* equilibrium state by simulating a time-evolution governed by the following combination of Glauber and Kawasaki dynamics (cf. Sec. 2.2.3): The sites occupied by rotors are visited randomly. For each rotor considered, a nearest-neighbor site on the triangular lattice is chosen at random. If this site is occupied, the rotor under consideration is rotated by a random angle between 0 and π, and the resulting configuration is the trial state. If, however, the neighboring site is vacant, the rotor and the vacancy are interchanged and the rotor simultaneously assigned a random angle between 0 and π, the resulting configuration then being the trial state. The difference in internal energy between the original state and the trial state is then calculated and the usual Monte Carlo acceptance criterion can be applied. By this construction, the mechanism by which the system relaxes to equilibrium is a combination of translational and rotational diffusion which obviously fulfils the requirements of ergodicity (cf. Sec. 2.2.2). It should be noted that the Glauber and Kawasaki dynamics are associated with a non-conserved quantity (orientational order) and a conserved quantity (particle number), respectively.

The quantities calculated include the internal energy per rotor $E(T) = <H>$ $/xN$, the specific heat per rotor $C(T) = (k_B T^2 xN)^{-1}(<H^2> - <H>^2)$, as well as the orientational pinwheel order parameters $\psi_\alpha(T)$ and the vacancy order parameters $\eta_\alpha(T)$, with $\alpha = 1, 2, 3, 4$. The pinwheel order parameters may be constructed from appropriate linear combinations of herringbone order parameters defined on the four sublattices of the pinwheel ground state in Fig. 5.3.3. The vacancy order parameters are defined as

$$\eta_\alpha = (xN)^{-1} < \sum_{i \in \omega_\alpha} (1 - x_i) > -(1 - x)/x, \qquad (5.3.10)$$

where ω_α, $\alpha = 1, 2, 3, 4$, denotes the four sublattices indicated in Fig. 5.3.3.

The simulations have been carried out for series of increasing as well as decreasing temperatures. We first discuss the results of heating scans where the system has been initiated in the pinwheel ground state. Secondly, the results of cooling scans will be presented. The temperature dependence of the internal energy per rotor, $E(T)$, is shown in Fig. 5.3.9. $E(T)$ is only slightly affected by the finite size of the lattice, as was also found to be the case for $x = 1$ in Fig. 5.3.4. Furthermore, $E(T)$ has an inflection point between $0.36K$ and $0.38K$.[*] The finite-size effects are much more pronounced for the order parameters given in Fig. 5.3.10. The actual order parameters plotted in this figure are the components of ψ_α and η_α in which the long-range order is found to reside. For $T \leq 0.35K$, all four lattice sizes give the same values of the order parameters. However, at higher temperatures pronounced finite-size rounding occurs, quite similar to that reported in Monte Carlo studies of the magnetization of the standard two-dimensional Ising ferromagnet (Landau 1976a). We note that both order parameters for all lattice sizes vary continuously through the transition region.

---

[*] For convenience, we have absorbed a factor k_B in the coupling constant K.

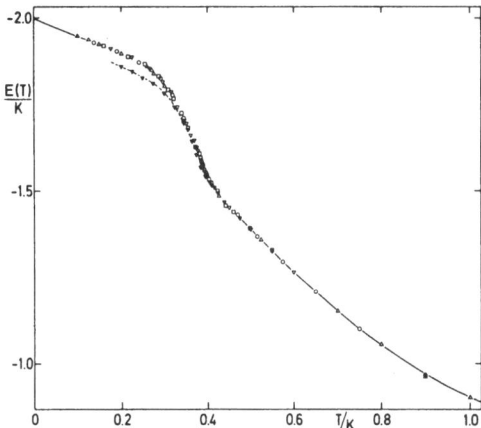

Fig. 5.3.9. Internal energy per planar rotor as a function of temperature for the annealed diluted anisotropic-planar rotor model at coverage $x = \frac{3}{4}$. The data derive from Monte Carlo calculations on lattices with $N = L^2$ sites. $L = 20$ (triangles), 28 (circles), 40 (inverse triangles for heating scans and solid inverse triangles for cooling scans), and 100 (squares).

The low-temperature behavior of the orientational order parameter is governed by gapless spin-wave excitations. To lowest order, spin-wave theory predicts (Harris et al. 1984)

$$\psi_\alpha(T) \simeq 1 - 0.160 T/K. \tag{5.3.11}$$

This is in excellent agreement with the Monte Carlo results in Fig. 5.3.10 where the data for $T \lesssim 0.22K$ may be constrained to a linear fit with slope -0.162 ± 0.003. The excitation spectrum for the vacancy order parameter has an energy gap corresponding to the energy which is spent in the process of moving a molecule into a pinwheel center followed by a relaxation of the involved neighboring molecules. The theoretical prediction (neglecting particle-hole bound states) is (Harris et al. 1984)

$$\eta_\alpha \simeq 1 - A e^{-\Delta E/T} \tag{5.3.12}$$

with an activation energy $\Delta E = (2 - \epsilon)K$, where $\epsilon = 0.15$ is the energy associated with the relaxation of neighboring molecules. It turns out that very long Monte Carlo runs are necessary to yield reliable estimates of this activation energy. This is probably due to the fact that whenever a molecule in a pinwheel structure at low temperature is moved into a pinwheel center, there is a very high probability that a later trial move will bring the molecule back into its old position. Therefore, a rather common excitation in the simulation process will be creation of particle-hole bound states which have a low affinity for dissociation. This explains why ordinary statistics involving 5000 - 10000 Monte Carlo steps per site for $T \lesssim 0.24$ leads to values of η_α, being unity within a 0.02% uncertainty. To get a better estimate of ΔE, some very long runs ($\sim$ 25000 excitations per site) were performed at $T = 0.238K$ and $0.278K$. These runs showed that, while ψ_α remained unchanged, η_α decreased very slowly. From the resulting data (which may not necessarily represent

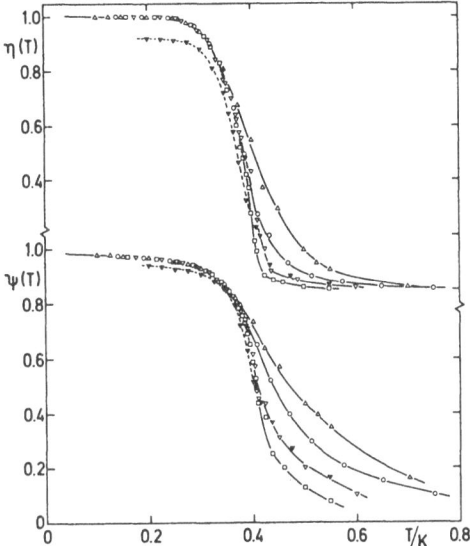

Fig. 5.3.10. Temperature dependence of the pinwheel order parameter, $\psi(T)$, and the vacancy order parameter, $\eta(T)$, for the annealed diluted anisotropic-planar rotor model at coverage $x = \frac{5}{4}$. The symbols are as in Fig. 5.3.9.

the true equilibrium state), we obtain a rough estimate of ΔE, $\Delta E/K \simeq 2.0 \pm 0.5$. This estimate is consistent with the theoretical prediction of $\Delta E/K = 1.85$. Still, it is not accurate enough to resolve the contribution from the relaxation of neighboring molecules. It is likely that a different Monte Carlo excitation scheme would be more efficient to determine the low-temperature properties of the present model (cf. e.g. Bortz et al. (1981) for an efficient scheme to obtain the low-temperature thermodynamics of Ising models).

The rapid decrease of both order parameters in a narrow temperature region is accompanied by a pronounced peak in the specific heat, $C(T)$, shown in Fig. 5.3.5. This very narrow and sharp peak signals a single phase transition. For the sake of clarity, Fig. 5.3.5 for $x = \frac{3}{4}$ only includes specific heat data for the $N = 10000$ lattice. A decrease in N slightly broadens the peak, moves the position of the peak maximum to lower temperatures, and decreases the peak intensity.

Before we discuss the nature of the transition as inferred from the observations presented above, we shall report on the Monte Carlo results for cooling scans where the system has been initiated in high-temperature orientationally and positionally disordered configurations. Results of such scans are also shown in Figs. 5.3.9 and 5.3.10 for the internal energy and the order parameters of a lattice with $N = 1600$ sites. For $T \gtrsim 0.35K$, the results are the same as those obtained from heating scans. However, at lower temperatures the results of the two sets of scans deviate increasingly as the temperature is lowered. Within our observation time ($\lesssim 8000$ Monte Carlo excitations per site), the results of the cooling scans relax only insignificantly towards the results of the heating scans. Even for the lowest temperature studied, we have not been able to bring the system back into a uniform pinwheel state. The cooling scan curves have the same sense of curvature as the heating scan curves. This dependence on the thermal history of the system is therefore not the usual phenomenon of hysteresis. We believe that the cause of this behavior is the extremely slow kinetics associated with the formation of ordered domains in systems with more order parameter components than spatial dimensions (Safran 1981, Sec. 5.4). Similar slow kinetics has been found in the case $x = 1$ and for several other models with highly degenerate order parameters (cf. Sec. 5.4). When the system is cooled below the transition temperature, a mosaic structure of ordered domains described by different order parameter components may develop. The annealing of this mosaic structure is extremely slow for topological reasons. The domain-growth kinetics of herringbone and pinwheel phases will be discussed in detail in Sec. 5.4 as an example of a numerical simulation of growth. In conclusion, we interpret the systematic deviation between the results from heating and cooling scans not as an ordinary hysteresis effect, but rather as due to the presence of long-lived non-equilibrium defects (domain walls) in ordered phases generated by cooling down from the disordered phase.

The conclusion of the discussion so far is that the model at coverage $x = \frac{3}{4}$ undergoes a single thermally induced phase transition which takes the pinwheel phase into an orientationally and positionally disordered phase. There are no signs, within the resolution of the numerical simulation, of a region of phase separation. The question now arises: what is the nature of the phase transition? In principle it is possible to provide reliable evidence that a phase transition is of first order by showing that it gives rise to discontinuities in first derivatives of the free energy (i.e. the order parameters) and that it is associated with the presence of metastable states and hysteresis (cf. Sec. 2.2.9). For a continuous transition it is, however, virtually impossible to prove conclusively in any experiment or computer simulation that the transition is not of first order. It can always be argued that a seemingly continuous transition may, in fact, be a »weakly« first-order one with unresolved small discontinuities. Furthermore, finite-size effects tend to smear small discontinuities and hence obscure weak first-order transitions. When dealing with possible continuous transitions, the best one can do is to analyze as many physical quantities as possible in terms of critical behavior and then compare the results to analyses of transitions of a well-established nature.

The continuous variation of the internal energy in Fig. 5.3.9 and of both order parameters in Fig. 5.3.10 for all values of N studied suggests a continuous transition. This continuous behavior should be contrasted to the discontinuous jump in the herringbone order parameter found for $x = 1$ (Fig. 5.3.7) when $N \gtrsim 6400$. The significant finite-size dependence of $C(T)$ for $x = \frac{3}{4}$ is in marked contrast to the lack of size dependence for $x = 1$. The specific heat curves for the two coverages are compared in Fig. 5.3.5. This figure reveals another striking difference between the two curves, namely that $C(T)$ for $x = \frac{3}{4}$ has a pronounced high-temperature tail which gives it a strongly asymmetric shape compared to the symmetric $C(T)$ for $x = 1$. Both specific heat curves converge at low temperatures to $k_B/2$ since the classical rotor models support spin-wave-like excitations down to zero temperature.

Taking the transition temperature of the finite lattice, $T_c(L)$, as the temperature where $C(T)$ attains its maximum value, we may estimate the transition temperature in the thermodynamic limit, $T_c = T_c(L \to \infty)$, using the finite-size scaling theory for systems showing critical behavior (Sec. 2.2.8). According to this theory, one has (Eq. (2.2.23))

$$T_c - T_c(L) \sim L^{-1/\nu}, \quad L \to \infty, \tag{5.3.13}$$

where ν is the critical exponent pertaining to the correlation length. In Fig. 5.3.11 is shown a finite-size scaling plot of $T_c(L)$ vs L^{-1}. Since the exponent ν is not known for the model studied here, we have assumed $\nu = 1$ without any loss of accuracy in the extrapolation to the infinite-lattice limit (cf. Fig. 3.1.5 for a similar scaling plot in the case of the three-dimensional Ising ferromagnet). Figure 5.3.11 shows that all four lattice sizes are well inside the asymptotic region described by finite-size scaling theory. A simple extrapolation then yields the critical temperature as $T_c/K = 0.400 \pm 0.008$. We conclude that the finite-size scaling properties of the specific heat are consistent with a continuous transition.

Having established the continuous nature of the transition, we proceed to derive critical exponents for the two order parameters as defined by the asymptotic power laws

$$\psi_\alpha(T) \sim (-t)^{\beta_\psi} \tag{5.3.14}$$

and

$$\eta_\alpha(T) \sim (-t)^{\beta_\eta}, \tag{5.3.15}$$

with $t = (T - T_c)/T_c \to 0$. The values of the exponents may be used to identify the universality class of the transition. Figure 5.3.12 is a log-log plot of the order-parameter data in Fig. 5.3.10. Due to finite-size rounding, data points for the smaller lattices lie consistently above those for the larger lattices as $-t$ decreases. In the decade, $0.02 \lesssim 0.2$, the data points for the larger lattices fall on common linear curves for both order parameters leading to the following estimates of the exponents in Eqs. (5.3.14) and (5.3.15)

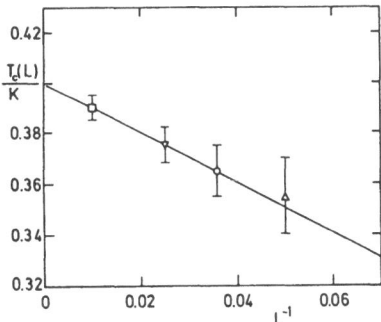

Fig. 5.3.11. Finite-size scaling plot showing the variation of the finite-lattice transition temperature, $T_c(L)$, with inverse linear lattice size, L^{-1}, cf. Eq. (5.3.13). $T_c(L)$ is estimated from the position of the maximum of the specific heat of lattices with $N = L^2$ sites.

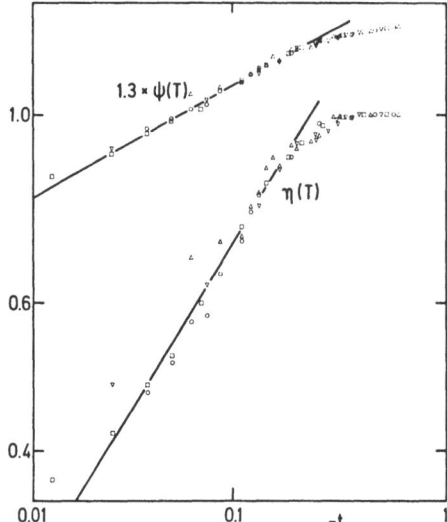

Fig. 5.3.12. Log-log plot of the order parameter data in Fig.5.3.10. The reduced temperatures, $t = (T - T_c)/T_c$, are obtained by use of the infinite-lattice transition temperature, $T_c = 0.400K$, derived from the finite-size scaling plot in Fig. 5.3.11. The solid lines correspond to simple power-law behavior, Eqs. (5.3.14) and (5.3.15), with exponents $\beta_\psi = 0.13$ and $\beta_\eta = 0.38$ for the pinwheel and vacancy order parameters, respectively. The symbols are as in Fig. 5.3.10.

$$\beta_\psi = 0.13 \pm 0.02, \quad \beta_\eta = 0.38 \pm 0.05. \qquad (5.3.16)$$

The major source of uncertainty in the exponents is the uncertainty in T_c. The value of $\beta_\eta/\beta_\psi = 2.9 \pm 0.6 \approx 3$ suggests that the vacancy order parameter is a non-critical order parameter. The mean-field predictions are $\beta_\psi = \frac{1}{2}$ and $\beta_\eta/\beta_\psi = 2$. Within mean-field (or Landau) theory, the relation $\beta_\eta = 2\beta_\psi$ is a consequence of a non-linear coupling of the two order parameters.

Our finding of a continuous transition of the pinwheel phase into the orientationally disordered phase is in agreement with the renormalization group calculations which predict that continuous transitions are allowed for systems in the universality class of the two-dimensional Heisenberg model with corner-type cubic anisotropy (sec. 5.3.3). Furthermore, the Monte Carlo value of the order parameter exponent, $\beta_\psi = 0.13 \pm 0.02$, is in excellent agreement with the renormalization group prediction of Ising critical behavior ($\beta = \frac{1}{8}$) for this universality class. Continuous transitions are also indicated by recent experimental investigations of the melting of molecular monolayers forming $p(2 \times 2)$ phases, such as oxygen on $Ni(111)$ (Roelofs et al. 1981) and freon an graphite (Kjaer et al. 1982), which are expected to display phase transitions in the same universality class (Schick 1981, Bak and Bohr 1983). Furthermore, the low-energy electron-diffraction experiments by Roelofs et al. allow critical exponents to be determined which are also found to be Ising-like (e.g. $\beta = 0.14 \pm 0.02$).

Finally, we shall discuss the complete $(x - T)$ phase diagram of the annealed diluted anisotropic-planar rotor model as it may be constructed on the basis of Monte Carlo, mean-field, and spin-wave calculations (Harris et al. 1984). Firstly, we note that the Monte Carlo calculations at coverage $x = \frac{3}{4}$ have established that the pinwheel structure remains stable at finite temperatures, that no phase separation is found, and that a single phase line separates the homogeneous pinwheel phase from the disordered phase. This is in disagreement with the mean-field theory which leads to the unusual prediction that the pinwheel phase is stable only along the herringbone-disordered state coexistence line in the $(\mu - T)$ diagram (μ is the chemical potential of the vacancies). This rather unphysical prediction is rooted in a remarkable symmetry in the mean-field theory which leads to a linear relation between the free energies of the three phases. This in turn implies that the pinwheel phase is only stable in coexistence with the two other phases. However, a spin-wave calculation shows that fluctuations break this symmetry and lead to a finite region of stability of the pinwheel phase around $x = \frac{3}{4}$.

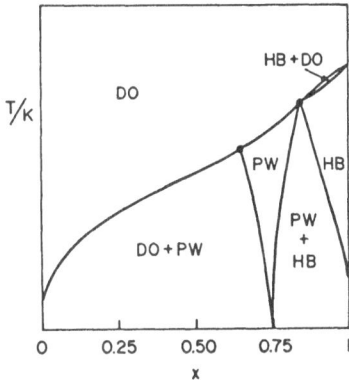

Fig. 5.3.13. Conjectured phase diagram $(x - T)$ for the annealed diluted anisotropic-planar rotor model. x is the concentration of rotors and T/K is the temperature.

This prediction is in accordance with the Monte Carlo calculations. On the basis of these considerations, the phase diagram in Fig. 5.3.13 has been conjectured. Obviously, more Monte Carlo work is called for in order to assess the validity of this diagram in regions away from $x = \frac{3}{4}$ and 1. To map out the complete phase diagram $(x - T)$ by Monte Carlo techniques would require reliable criteria for locating phase coexistence regions. Since it is rather difficult to use such criteria unambiguously, it would be more convenient to perform the computer simulations within the grand canonical ensemble where μ rather than x is the independent thermodynamic variable. In the $(\mu - T)$ phase diagram, the coexistence regions of Fig. 5.3.13 appear as simple phase lines.

5.3.7   Physical realizations of the anisotropic-planar rotor model with vacancies

N_2 on graphite belongs to the island-forming adsorbate systems (Lagally et al. 1978) which at submonolayer coverages, $x < 1$, form positionally ordered regions rather than a random lattice gas. Indeed, the calorimetric measurements by Migone et al. (1983) at submonolayer coverages, $x = 0.87$ and 0.29, show that the specific heat peak remains unchanged for $x \leq 1$ indicating that the N_2 molecules undergo the same cooperative process (the herringbone transition) independent of $x(\leq 1)$. Thus, N_2 at submonolayer coverages is not a physical realization of the annealed diluted anisotropic-planar rotor model.

A possible realization is a mixture of N_2 and rotationally inert gas atoms, such as the rare gases Ar and Kr. The rare gas atoms, with a cross-sectional area not much different from that of N_2, would then play the role of the vacancies and stabilize the pinwheel structure by placing themselves at the centers of the pinwheels. Preliminary studies of N_2-rare gas mixtures performed by Satija and Passell[*] show, however, that there is a tendency for the rare gas atoms to phase separate. The molar fraction of rare gas atoms which can be dissolved in the N_2 system appears to be much less than $\frac{1}{4}$. These observations suggest that the direct vacancy-vacancy interactions left out of the diluted rotor model may play an important role.

A mixture of para-H_2 and ortho-H_2 would be another possible candidate for a physical realization of the diluted anisotropic-planar rotor model, in the sense that the spin-1 nuclear quantum quadrupoles of ortho-H_2 would act as rotors (Harris and Berlinsky 1979, Kubik and Hardy 1978) and the para-H_2 molecules would not interact anisotropically. It is not clear, though, whether the sticking coefficient of the H_2 molecules is low enough to allow the system to anneal by diffusion. Also, recent nuclear magnetic resonance experiments on ortho-H_2 on graphite seem to have detected a transition into the pinwheel phase (Kubik et al. 1984). Thus, for ortho-H_2 the pinwheel phase may be stable even close to $x = 1$ indicating that the substrate potential may be positive (Fig. 5.3.2) and the anisotropic-planar rotor model therefore not an appropriate model.

5.4 Kinetics of Growth

5.4.1   Growth

Formation and growth of aggregates and ordered structures from the microscopic constituents of matter are widely observed processes in nature. Examples include grain growth during crystallization, gelation of polymers, blood coagulation, flocculation in colloids, antigen-antibody reactions, and percolation and nucleation in phase transitions and critical phenomena. The interest in growth processes is

---

[*] S. Satija and L. Passell, preliminary experiments.

fairly old and the technological importance of such processes is well known from metallurgy (see e.g. Carpenter and Elam 1920, Haessner 1978).

Recently, there has been a renewed interest in the kinetics of growth processes, theoretically as well as experimentally. It has turned out, however, to be extremely difficult to conduct quantitative experiments designed to measure the kinetics of growth. Moreover, the theory of growth is only in its very early stage of development. An understanding of growth kinetics is just beginning to emerge.

Despite the diversity of growth processes, a unifying principle seems to be operative. This principle is related to scale invariance. The origin and importance of this scale invariance are perhaps best illustrated by considering a condensed system which, in equilibrium, will form a uniform ordered structure at low temperatures. Suppose the system is prepared in its disordered state at high temperatures and then suddenly quenched to low temperatures. An ordering process will then take place by the simultaneous formation and growth of a large number of droplets or domains which represent local regions with a high degree of ordering. The average radius, $R(t)$, of the domains will increase in time and the domains will eventually coarsen in order to reduce the excess free energy associated with the domain walls. Finally, the ordering of a single domain will take over and extend to the entire system which then has reached its new state of thermodynamic equilibrium. In the later stages of the growth, $R(t)$ will exceed the characteristic microscopic length scales of the system, e.g. as given by the nearest-neighbor distance, r_0. When $R(t) \gg r_0$, the microscopic details of the system are expected to become irrelevant and a universal behavior should result. $R(t)$ is then the only relevant length scale of the problem.

Binder and Stauffer (1974) were the first to point out that these ideas of dynamical scaling may be expressed as a homogeneity relation for the structure factor, $S(\bar{q}, t)$,

$$S(\bar{q}, t) = R^d(t) F[\bar{q}R(t)], \qquad (5.4.1)$$

where d is the spatial dimension and F is an (unknown) scaling function. From Eq. (5.4.1) it follows that the domain growth is algebraic in time and is governed by the growth law

$$R(t) \sim t^a, \qquad (5.4.2)$$

where a is a universal growth exponent. The analogy with static scaling within the theory of critical phenomena is obvious (Sec. 2.1.2). This analogy also suggests that the various growth processes may be characterized by a small number of universality classes, each class being determined by only very general properties of the system under consideration. The sorting out of the different universality classes is far from complete. * However, important parameters to specify the universality classes seem to be the degeneracy of the ordered state and the type of conservation laws governing the dynamics.

In the theoretical description of domain-growth kinetics and in the calculation of the growth laws, Eq. (5.4.2), we are facing the very fundamental and difficult

---

* Furukawa (1984) has recently questioned the universal behavior of domain-growth kinetics.

problem of systems far from equilibrium (Furukawa 1984). This is a highly non-linear problem in which the order parameter fluctuations play an important role. Related problems are dynamics of random interfaces (Grant and Gunton 1983, Kaski et al. 1983b) and the kinetics of phase separation (nucleation and spinodal decomposition) (see e.g. Gunton et al. 1983 for a recent review).

Experimental investigation of domain-growth kinetics is hampered by serious restrictions in time-resolution. The classical method in metallurgy proceeds via electron microscopy (Allen and Cahn 1979). Modern low-angle scattering techniques look promising, in particular when synchrotron sources are applied. Low-energy electron-diffraction studies have proved useful to study the domain-growth kinetics of chemisorbed overlayers (Lagally et al. 1978, Wang and Lu 1983).

The kinetics of domain growth in systems undergoing crystallization is expected to depend upon the relation between the number, p, of thermodynamically degenerate ordered phases and the spatial dimension, d, of the system. Building on some early ideas proposed by Lifshitz (1962), Safran (1981) has recently advanced a time-dependent Ginzburg-Landau theory which predicts that the domain size at low temperatures equilibrates as a power-law for $p < d + 1$ and as a logarithmic function of time for $p \geq d + 1$. The basic idea behind this prediction is that for $p < d + 1$, the main driving force for the growth is the curvature of the domain interfaces (Allen and Cahn 1979). For $p \geq d + 1$, the curvature-driven forces become less effective due to a competition with the pinning effects dictated by the topology (Safran 1981, Safran et al. 1983).

For $p < d + 1$, power-law behavior has indeed been observed in experiments on binary alloys ($d = 3$, $p = 2$) (Allen and Cahn 1979) and on oxygen adsorbed on $W(122)$ surfaces ($d = 2$, $p = 2$) (Wang and Lu 1983). For $p \geq d + 1$, very slow kinetics has been encountered in experiments on grain growth in polycrystalline metals ($d = 3$, $p \to \infty$) (Gordon and El-Basyoumi 1965) and on the ordering of oxygen on $W(110)$ surfaces ($d = 2$, $p = 4$) (Lagally et al. 1978).

5.4.2   Computer simulation of domain-growth kinetics

Domain-growth kinetics is expected to be strongly influenced by impurities (Haessner 1978) and by surface inhomogeneities in the case of adsorbed overlayers. Hence, it is difficult to interpret experimental kinetic data within a universal classification scheme. Model studies are therefore of particular importance in scrutinizing the theoretical ideas. At present, computer simulation is essentially the only method which allows a quantitative study of the domain-growth kinetics of microscopic models. A number of Monte Carlo studies focussing on growth kinetics have appeared within the last few years using lattice models, e.g. simple Ising antiferromagnets ($d = 2$, $p = 2$) (Phani et al. 1980b, Sahni et al. 1981), ($d = 2$, $p = 4$) (Sahni and Gunton 1981, Sacco and Chalupa 1981, Sadiq and Binder 1983, 1984), clock models ($d = 2$, $p = 6, 26$) (Kaski and Gunton 1983), and q-state Potts models ($d = 2$, $p \geq 2$) (Sahni et al. 1983a,b, Srolovitz et al. 1983). Furthermore, several simulation studies of the phase transitions in different models with $p > d$ have revealed extremely slow domain-growth kinetics which seems to be associated with the formation of metastable glass-like states (Mouritsen et al. 1977 (cf. Sec. 4.1.3), Banavar et al. 1980, Mouritsen and Berlinsky 1982).

The numerical simulation studies of lattice models show concordantly for $p < d + 1$ that the domain-growth kinetics obeys dynamical scaling and that algebraic growth laws apply. For $p \geq d + 1$, the results are more difficult to reconcile, partly because temperature and lattice effects may be important. In no case has logarithmically slow growth been observed although many models display algebraic growth with low exponents (cf. Secs. 5.4.3 and 5.4.4).

There are a number of difficulties associated with computer simulation studies of growth. First of all, the growth may be inhibited by an inefficient excitation mechanism.* Secondly, crossover to critical relaxation may influence the effective growth exponents (Safran et al. 1982, Sahni et al. 1983c, Sadiq and Binder 1984). Finally, in most cases there is not sufficient experimental information available to relate the characteristic time step of the Monte Carlo procedure to real time.

5.4.3   Domain-growth kinetics of herringbone phases

Our example of a computer simulation study of domain-growth kinetics comes from the field of statistical surface physics. In Sec. 5.3, we studied an anisotropic-planar rotor model on a triangular lattice as a model of nitrogen molecules physisorbed on graphite (Eq. (5.3.1) with $J = 0$). At coverage $x = 1$ and at low temperatures, this model gives rise to the orientationally ordered herringbone structures in Fig.5.3.1. The herringbone orientational order parameter is three-fold degenerate. Since each of the three herringbone structures can be embedded into the triangular lattice in two different ways (connected by a simple translation along the propagation vector), the number of thermodynamically degenerate ordered domains is $p = 6$. Thus, by the anisotropic-planar rotor model on a triangular lattice, we have at hand a model with $p > d$ which is expected to display interesting domain-growth kinetics.

The domain-growth kinetics of the herringbone phases has been studied using computer simulation techniques by Mouritsen (1983b). This is the first computer study of the domain-growth kinetics in a realistic microscopic model of a specific physisorbed system. It is important to note that the Hamiltonian of this system not only contains the correct symmetry of the physical system, but also includes the appropriate energetics (cf. Sec. 5.3.2). The computer simulation thus provides experimental information on a system which is difficult to study by conventional experimental techniques since contamination and surface inhomogeneities are likely to influence the kinetics. Other computer simulation studies of specific adsorbed monolayer systems have focussed only on reproducing the symmetry of the ordered domains (see e.g. Sahni and Gunton (1981) for a study of the kinetics of island growth in a model of the $p(2 \times 1)$ phases of oxygen chemisorbed on

---

* E.g. the use of Glauber dynamics to anneal a simple Ising antiferromagnet ($d = 2$, $p = 2$) in quenches from $T \gg T_c$ to $T \lesssim 0.3 T_c$ will not be successful but rather rapidly lead to a completely frozen-in glas-like domain structure.

W(110)). Moreover, these other studies have employed models with discrete degrees of freedom (e.g. Ising and Potts models). The continuous single-site rotor variables of the anisotropic-rotor model introduce additional dynamical features which may influence the domain-wall formation and the growth kinetics. In the first computer study of the anisotropic-planar rotor model, emphasis was on determining the growth law governing the herringbone domains (Mouritsen 1983b). Kaski et al. (1984) has later supplemented this study by investigating the scaling properties of the structure factor. Below, we shall describe the combined results of these two studies to illustrate the potentials in computer simulations of domain growth.

We want to study the time evolution of a triangular array of rotors governed by the anisotropic-planar rotor model, Eq. (5.3.1), as it develops after a deep quench in temperature from $T \simeq \infty$ to $T \simeq 0$. Immediately after the quench, the system is far from equilibrium and the true dynamical evolution following the quench can only be obtained by solving the complete set of equations of motions, $\partial \varphi_j(t)/\partial t = [\varphi_j(t), H]$, e.g. as it may be done numerically in a molecular dynamics calculation. However, this is a very complicated procedure which may be extremely time consuming since a large number of time steps is expected to be necessary in order to follow the system over a reasonable time span. Rather, we shall exploit the dynamical process inherent in the usual Monte Carlo method (cf. Sec. 2.2.2) to construct the time evolution. This allows for a suitable time scale of the kinetics at the expense of losing the precise relationship between Markov time and real time. This is not a serious restriction as long as we are mainly interested in the general aspects of domain growth. The time, t, after the quench is conveniently measured in units of Monte Carlo steps per site.

The Monte Carlo excitation mechanism used for the present model is a single-site Glauber-type excitation mechanism which takes the system from state α to state β with a probability $p_{\alpha\beta} = p_{\alpha\beta}^* \exp[-(E_\beta - E_\alpha)/k_B T]$ (cf. Eqs. (2.2.9) and (2.2.10)). The ergodic stochastic matrix $\overline{\overline{P}}^*$ basically sets the time scale of the problem, $p_{\alpha\beta}^* \sim \tau^{-1}$. In order to examine the dependence of the domain growth upon details of the local excitation mechanism, $p_{\alpha\beta}^*$ is chosen in two different ways corresponding to random reorientations (rotational diffusion) of the rotor angle, $\varphi_i \rightarrow \varphi_i + \Delta\varphi_i$, whithout (a) $0 \leq \Delta\varphi_i < \pi$ and with (b) $0 \leq \Delta\varphi_i \leq \pi/5$ an angular restriction. The choice of Glauber dynamics implies that the order parameter is not a conserved quantity during the relaxation towards equilibrium. The lattice sites are visited sequentially. Due to the continuous nature of the rotor variables, we have not found it necessary to use special high-efficiency excitation mechanisms which have recently been used to study deep quenching for discrete Ising and Potts models (Sahni et al. 1983b, Sadiq and Binder 1984, Sadiq 1984). By these mechanisms, basically only the particles in the domain boundaries are excited.

The present simulations are carried out on triangular lattices subjected to toroidal periodic boundary conditions. To reduce possible boundary effects, the main results are obtained for a very large lattice with $N = L \times L = 152 \times 152$ sites. Also, in the low-t regime some results are reported for a 40×40 lattice. Kaski et al.'s results are derived from a lattice with 60×60 sites. The initial configuration of rotors characteristic of high temperatures is chosen as a random configuration. The quench is to temperatures typically around 0.02 - 0.05 K/k_B.

Internal energy, order parameters, as well as the microscopic configuration of rotors are recorded as a function of t. In order to provide good statistics, the results are averaged over several quenches corresponding to different initial high-temperature configurations as well as different random number sequences. The evolution of the large lattice is followed for times up to $t = 6000$. The 40×40 lattice displays finite-size effects around $t = 600$. Kaski et al. followed the 60×60 lattice up to around $t = 4500$ where finite-size effects are seen to set in by comparing with the 152×152 lattice. Presumably, the largest lattice could be followed to even higher values of t without significant finite-size rounding. Finite-size effects become important, just as they do in critical phenomena, when the size, $R(t)$, of highly correlated regions becomes comparable to the size, L, of the system. In the case of deep quenching to temperatures far below the transition temperature where fluctuations and roughening are unimportant, a highly correlated region is simply an ordered domain characterized by a high value (≈ 1) of one of the order parameter components. When $R(t)$ becomes comparable to L, or when a single domain has formed a percolated cluster, the domain growth is expected to reflect the finite size of the system.

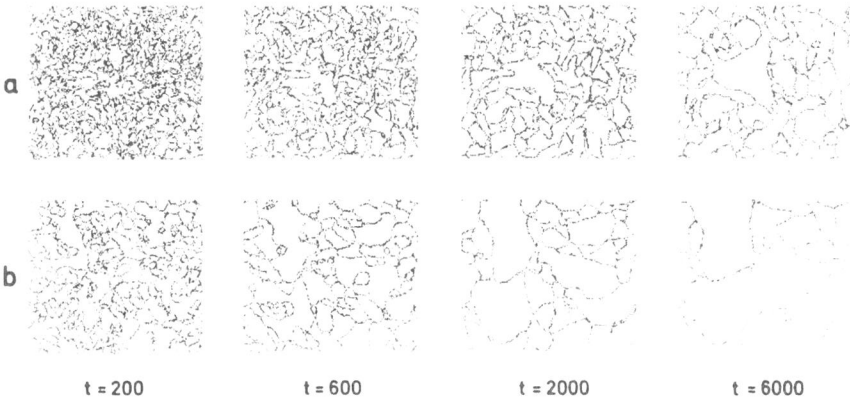

a

b

| t = 200 | t = 600 | t = 2000 | t = 6000 |

Fig. 5.4.1. Snapshots showing the distribution of herringbone domains at various times, t [Monte Carlo steps per site], following a quench in temperature from $T \simeq \infty$ to $T \simeq 0$. The system contains 152×152 sites. Only the domain walls are displayed. Panels (a) and (b) correspond to unrestricted and restricted single-site excitation mechanisms, respectively.

In Fig. 5.4.1, snapshots are given of the configurational state of the large system for a series of selected times. For clarity, only the domain walls are displayed. In a lattice model with continuous single-site variables, the walls are »soft«, in contrast to the »hard« walls of Ising and Potts models, and the extension of the walls is a matter of definition. Here, we have chosen to consider rotors as part of a domain wall if their value of φ_i deviates more than $\pi/15$ from the ground state values of the adjacent ordered domains. The value of this discrimination angle, which only influences the thickness of the wall and not its position, is of marginal importance at low temperatures. However, at higher temperatures and especially near the phase transition, where roughening fluctuations are important, the criterion for locating the domain walls has to be reconsidered. Figure 5.4.1 shows that for small t, the cluster distribution is very ramified and the shapes of the clusters are irregular. The regularity in shape increases with t. No significant decrease in domain wall thickness with time is observed for $t \gtrsim 50$. For the range of times studied here, no general form has been found for the internal structure of the walls. Neither do there seem to be any preferred directions of the walls relative to the lattice. However, for the smaller lattice, there is some indication that the optimal directions are the three canonical axes of the triangular lattice and that the pinwheel is a typical wall configuration. The two panels in Fig. 5.4.1 show that the excitation mechanism based on a restricted reorientation angle facilitates formation of larger clusters. When two domains of the same type of ordering meet, they coalesce. The coalescence processes occur most frequently for small t. From visual inspection of snapshots such as those presented in Fig. 5.4.1, some general conclusions on domain growth for large t can be drawn: (i) the domains change size by migration of the walls, (ii) the driving force of the migration is the curvature of the walls, and (iii) the points where three domains meet act as pinning centers (Sahni et al. 1983a,b), the effectiveness of the pinning being commensurate with the similarity of the angles at which the domain walls meet. These observations are in line with results found in classical work on recrystallization phenomena (Carpenter and Elam 1920, Burke and Turnbull 1952) and in model studies of q-state ($q \geq 3$) triangular Potts models (Sahni et al. 1983a,b).

A quantitative analysis of the domain growth requires definition of a length scale. That only a single length scale is important for the present problem has been demonstrated by Kaski et al. (1984) through their analysis of the dynamical scaling properties of the growth process in Fig. 5.4.1. These authors calculate the three components of the circularly averaged dynamical structure factor (Marro et al. 1975)

$$S_\alpha(\overline{q}, t) = N^{-1} < | \sum_{j=1}^{N} \sin(2\varphi_i - 2\theta_\alpha)e^{i(\overline{q} + \overline{q}_\alpha)\cdot \overline{r}_j}|^2 >, \qquad (5.4.3)$$

where $\alpha = 1, 2, 3$. The directional angles, θ_α, and the propagation vectors, $\overline{q}_\alpha$, of the herringbone structures have been defined in Eq. (5.3.4). The moments of the average structure factor, $S(\overline{q}, t) = \sum_\alpha S_\alpha(\overline{q}, t)$,[*] are defined as

$$k_n(t) = \sum_{\overline{q}} |\overline{q}|^n S(\overline{q}, t) / \sum_{\overline{q}} S(\overline{q}, t). \qquad (5.4.4)$$

Kaski et al. then proceed by constructing the scaling function, F, of the dynamical structure factor, Eq. (5.4.1),

$$F(\overline{x}) = k_2(t)S(\overline{q}, t), \quad \overline{x} = \overline{q}/\sqrt{k_2(t)}. \qquad (5.4.5)$$

---

[*] The sum of the three components of the structure factor is the experimentally accessible quantity in a scattering experiment.

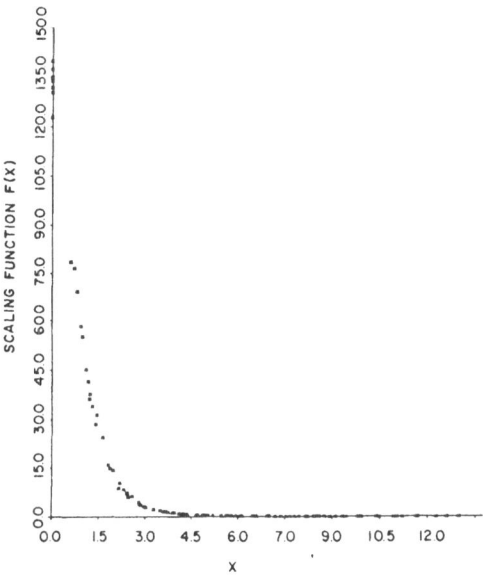

Fig. 5.4.2. The scaling function, $F(x)$ Eq. (5.4.5), for the dynamical structure factor in Eq. (5.4.3). [From Kaski et al. (1984) by courtesy of J.D. Gunton.]

The scaling function, $F(x)$, is shown in Fig. 5.4.2. This figure clearly shows that the growth of the herringbone phases obeys dynamical scaling characterized by a single time-dependent length scale. That scaling holds implies that the growth is a *self-similar* process.

Dynamical scaling suggests $k_2^{-\frac{1}{2}}(t)$ as an appropriate length scale. Another choice of length scale immediately suggests itself from Fig. 5.4.1, namely the average domain radius, $R(t)$. $R(t)$ may be calculated from the instantaneous domain distribution by defining the radius of a domain as the squareroot of the number of sites in the domain. A third choice of length scale in quenched systems with degenerate order parameters has been proposed by Sadiq and Binder (1983, 1984) in terms of the time-dependent root-mean-square order parameter

$$L(t) = [N \sum_{\alpha} \psi_\alpha^2(t)]^{\frac{1}{2}}/\psi(T). \qquad (5.4.6)$$

$\psi(T) \approx 1$ is the equilibrium value of the order parameter. Kaski et al. also calculate $L(t)$.

The results for $R(t)$ and $R^2(t)$ are shown in Fig. 5.4.3 as obtained from the two different studies (Mouritsen 1983b, Kaski et al. 1984). An excellent agreement is observed for $t \lesssim 4500$. Above $t \simeq 4500$, the results from the 60×60 lattice display finite-size effects. Figure 5.4.3 shows that the growth is very slow. The same slow growth is observed in terms of the other length variables, $k_2^{-\frac{1}{2}}(t)$ and $L(t)$ (Kaski et al. 1984).

We now proceed to determine the growth law, Eq. (5.4.2), written more generally as

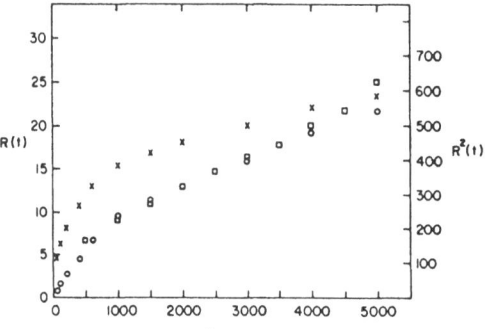

Fig. 5.4.3. Average domain radius, $R(t)$, and average domain area, $R^2(t)$, as functions of time, t [Monte Carlo steps per site]. The results of Kaski et al. (1984) ($R^2(t)$: squares) and of Mouritsen (1983b) ($R(t)$: crosses, $R^2(t)$: circles) are compared in the case of the unrestricted local excitation mechanism (a). The data in the two studies derive from lattices with 60×60 and 152×152 sites, respectively. [From Kaski et al. (1984) by courtesy of J.D. Gunton.]

$$R(t) - R(t_o) = At^a. \tag{5.4.7}$$

This is rather troublesome since the numerical data cannot sustain the full three-parameter fit implied by Eq. (5.4.7). A basic problem is that the growth law may only be effective after a transient period, $t_o > 0$, and that therefore $R(t_o) \neq 0$. The transient period would characterize the stage of the growth process in which the herringbone domains are being formed. After this transient period, the domains will either grow or shrink. As pointed out above, the continuous rotor variables preclude an unambiguous definition of a domain wall and its extension. This is particularly true in the early stages of the process where basically the whole system consists of soft walls. Lacking sufficient information to determine t_o, two alternative ways of analyzing the data in Fig. 5.4.3 have been proposed.

The first analysis is built on the assumption that $R(t > t_o) \gg R(t_o) \approx 0$ (Mouritsen 1983b). From the log-log plot of $R(t)$ vs t in Fig. 5.4.4, the growth law may then be examined. For both excitation mechanisms, (a) and (b), this figure suggests that there are two time regimes, an initial fast one separated from a slower late-time behavior by a crossover region. The position and extension of the crossover region depend on the excitation mechanism. The most important conclusions to be drawn from Fig. 5.4.4 are that the growth is algebraic with time in both time regimes and that the associated growth exponents are the same for both excitation mechanisms. In the early regime, $a \simeq 0.40$ and in the late regime, $a \simeq 0.25$. These results suggest that a is universal and independent of details of the local excitation mechanism. In contrast, the prefactor A, Eq. (5.4.7), is a function of the details of the excitation mechanism and its intrinsic characteristic time scale. If the assumption underlying this analysis holds, we are thus faced with a growth exponent in the late-time regime which is dramatically lower that the classical value (Lifshitz 1962, Safran 1981), $a = \frac{1}{2}$, found for binary alloys (Allan and Cahn 1979) and Ising models with a non-conserved order parameter (Phani et al. 1980b, Sahni et al. 1981, Sadiq and Binder 1983, 1984). Moreover, it is

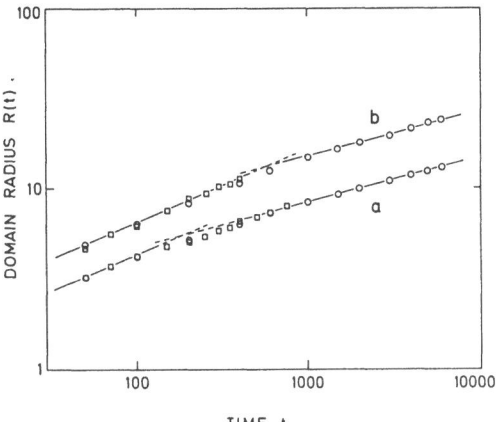

Fig. 5.4.4. Log-log plot of the average domain radius, $R(t)$, vs time, t [Monte Carlo steps per site]. Results are shown for two different local excitation mechanisms: unrestricted (*a*) and restricted (*b*) reorientation angles. Data are given for systems with 40×40 sites (squares) and 152×152 sites (circles).

smaller than *a* for any q-state Potts model, for which *a* decreases from $a = 0.495$ for $q = 3$ to $a = 0.41$ for $q \geq 26$ (Sahni et al. 1983a,b, Srolovitz et al. 1983). The much slower kinetics of domain growth found under the present assumption is probably caused by the continuous nature of the site variable φ_i, which makes it possible for the system to form soft domain walls. The softness of the walls screens the interaction between the different domains and decreases the driving force for the growth. It appears that a possible explanation of the presence of a separate early fast time regime is that the high frequency of coalescence processes at low t to some extent outbalances the slow kinetics caused by the screening of the domain-domain interactions.

The second type of analysis has been performed by Kaski et al. (1984) who question the above interpretation of the growth data. These authors base their analysis on the existence of a transient period. Assuming $t_o \simeq 640$, they show that the data (using any of the three length scale definitions) for $t > t_o$ can be constrained to a classical (Allen-Cahn) growth law with exponent $a = 0.50 \pm 0.03$. The resulting fit, which is shown for the average domain area, $L^2(t)$, in Fig. 5.4.5, is very good for all three choices of length scale. Thus, Kaski et al. conclude that the growth of herringbone phases is likely to be described by the classical Allen-Cahn theory for curvature-driven growth and that the softness of the rotor system only influences the non-universal parameters of the growth law, making the evolution of the domains very slow.

At present, it seems not possible to decide which of the two interpretations presented above is the correct one. However, both interpretations suggest concordantly that Safran's (1981) prediction of logarithmic growth, $R(t) \sim \ln t$, for systems with $p \geq d$ (Sec. 5.4.1) is not valid for the herringbone phases. This seems reasonable since Safran's theory assumes a continuum and presupposes that the equilibrated domains are ideal six-sided hexagons. The present system is a discrete

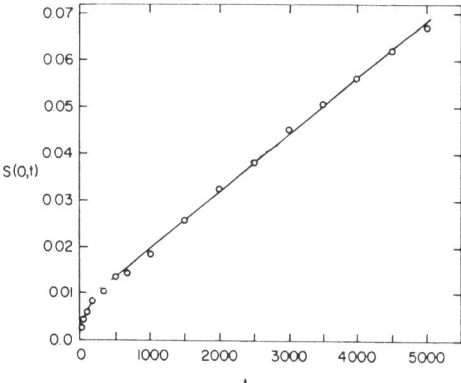

Fig. 5.4.5. Normalized average domain area, $S(0, t) = L^2(t)/N$, as a function of time, t [Monte Carlo steps per site]. $L(t)$ is the average linear extension of the domains as defined by Eq. (5.4.6). The solid line denotes the linear Allen-Cahn prediction with exponent $2a = 1$. [From Kaski et al. (1984) by courtesy of J.D. Gunton].

lattice model, and as Sahni et al. (1983a) have pointed out, domain wall defects on a triangular lattice are easily transmitted through the pinning centers leading to a more rapid migration of walls.   Also, it is obvious from Fig. 5.4.1 that odd-sided irregular domains frequently occur, which again facilitates the overall relaxation of the system (Carpenter and Elam 1920). So far, logarithmically slow growth has not been observed in any lattice model with $p \geq d$, although in several cases very low values of the growth exponent have been reported (Sacco and Chalupa 1981, Sahni and Gunton 1981). A recent computer simulation of the domain-growth kinetics in a non-lattice model of Kr adsorbed on graphite (Knak Jensen 1984) has, however, yielded a growth rate consistent with Safran's theory. It should be pointed out that Furukawa (1984) has recently argued that the growth laws need not be universal at all and that $p > d$ need not imply logarithmic growth.

The growth exponents extracted from the above analyses are found to be independent of details of the local excitation mechanism (rotational diffusion) in terms of which the domain growth proceeds. The predictions from the computer simulation studies of the domain-growth kinetics of herringbone phases may therefore be tested experimentally on monolayers of N_2 on graphite. A low-energy electron-diffraction experiment measuring the intensity of the superlattice peak as a function of time has been carried out for oxygen on $W(112)$ (Wang and Lu 1983). This intensity is directly proportional to $L^2(t)$ in Eq. (5.4.6). A similar experiment should be possible for N_2 on graphite. An extremely slow relaxation following quenches through the herringbone transition has recently been reported from a nuclear magnetic resonance study of $^{15}N_2$ physisorbed on graphite (Sullivan and Vaissiere 1983). This was attributed to the high degeneracy of the herringbone ordering.

Finally, it should be mentioned that herringbone ordering in liquid crystals (Bruinsma and Aeppli 1982) may obey similar growth kinetics. However, since the ordering in smectics is governed by a close-packing condition (Doucet 1979) rather than by electric quadrupole-quadrupole interactions, the internal structure of the domain walls is expected to be different.

5.4.4 Domain-growth kinetics of pinwheel phases

The presence of impurities or vacancies in a system undergoing crystallization is expected to influence the kinetics of growth (see e.g. Haessner 1978). The experimental finding of very low growth exponents for solid mixtures quenched into the miscibility gap (Hennion et al. 1982, Blaschko et al. 1982) indicates that the growth rate may be suppressed in the presence of impurities.

In this section, we shall briefly report on the first computer simulation study[*] carried out to investigate the domain-growth kinetics of a microscopic model with impurities. The results show that the impurities inhibit the growth rather dramatically. The model is the diluted anisotropic-planar rotor model, Eq. (5.3.5), which in Sec. 5.3.7 was proposed as a model of mixtures of N_2 and rare gases physisorbed on graphite. The vacancies in this model play the role of rotationally inert impurities which do not directly couple to the order parameter. For an impurity concentration $1 - x = \frac{1}{4}$, the equilibrium ground state is the eight-fold degenerate ($p = 8$) pinwheel structure shown in Fig. 5.3.3.

By use of Monte Carlo techniques similar to those described in Sec. 5.4.3, the evolution following quenches from high to low temperatures has been constructed. In all quenches performed, the growth is found to become extremely slow after an initial period of about $t = 30$ Monte Carlo steps per rotor. Visual inspection of snapshots of the microconfigurations at times $t \gtrsim 100$ shows that a rather complex domain pattern has developed. Domains described by the eight types of pinwheel ordering occur frozen-in together with domains of herringbone ordering. Precipitation of vacancies are also found which compensates for the denser herringbone domains. Thus it appears that not only do we have competition between eight different types of pinwheel ordering, but also competition with and among transient herringbone structures which have the same bulk domain energy per site ($-2K$) as the pinwheel structures.

To provide a quantitative analysis of this very complex behavior, we have found it useful to study the excess internal energy associated with the entire structure of domain walls. The excess internal energy per rotor, $\Delta E(t)$, is defined as

$$\Delta E(t) = E(t) - E(T), \qquad (5.4.8)$$

where $E(T)$ is the equilibrium internal energy at the temperature to which the system is quenched. In our case, $E(T) \simeq E(T = 0) = -2K$. ΔE is an approximate measure of the total domain wall energy in the later stages of the growth. According to an argument advanced by Binder and Stauffer (1974), $\Delta E(t)$ is expected to display a power-law behavior

$$\Delta E(t) \sim t^{-b}, \qquad (5.4.9)$$

---

[*] O.G. Mouritsen, unpublished calculations.

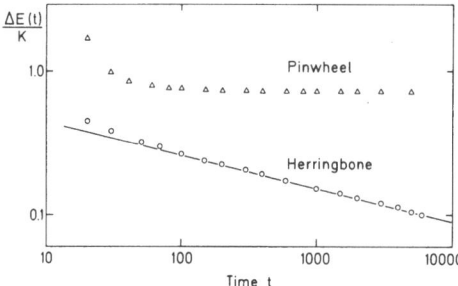

Fig. 5.4.6 Excess internal energy per rotor, $\Delta E(t)$ Eq. (5.4.8), plotted *vs* time, t, for the anisotropic-planar rotor model. Results are shown for the pure model ($x = 1$), which has a herringbone structure as its ground state, and for the impure model ($x = \frac{3}{4}$) which has a pinwheel structure as its ground state.

where the scaling relation $b = a$ (cf. Eq. (5.4.2)) holds far away from the critical point (Sadiq and Binder 1984).

In Fig. 5.4.6 is plotted $\Delta E(t)$ *vs* t for a system with 152×152 sites followed for times up to 5000 Monte Carlo steps per rotor. For comparison, we have in this figure also plotted $\Delta E(t)$ in the case of a pure system ($x = 1$, Sec. 5.4.3). The vacancies are seen to have a dramatic effect on the growth rate. For the impure system, the growth data cannot be described by the power law, Eq. (5.4.9). For the pure system, the algebraic growth law, Eq. (5.4.9), is found to describe the growth for times $t \gtrsim 50$ with a growth exponent $b \simeq 0.24$. Thus, in this case the scaling relation, $a = b$, is found to hold provided that the assumption, $R(t_o) \approx 0$, used in the previous section remains valid.

5.4.5 Kinetics of growth and critical phenomena

In concluding this section, it may be elucidating to draw the reader's attention to the stage of critical phenomena as it was set before the advent of the renormalization group theory more than ten years ago. At that time, no consistent framework existed within which the concept of universality could be rationalized. The bulk of experimental data and of theoretical calculations indicated, by the clustering of the critical exponents around a few distinct values, that some unifying concept might exist. Still, the exponent values for different systems were not completely the same. Questions then arised as to how »close« to a critical point one should go before the asymptotic critical behavior would manifest itself. It was the merit of the renormalization group theory to show that variables irrelevant at the critical point, T_c, may become important away from T_c and lead to correction-to-scaling and crossover between different kinds of critical behavior (Wegner 1972). With these new possibilities, the major part of the existing data on critical phenomena could be reconciled with the concept of universality.

In the present author's opinion, we are now witnessing a parallel development in the field of growth kinetics. As was the case for critical phenomena in the 1960s, data for growth kinetics are now being compiled and subjected to simple analyses.

Assuming scaling without corrections, effective exponents are derived in order to sort out the relevant parameters for a classification. So far, no unifying theory has been proposed although it has been speculated that the renormalization group ideas may to some extent be carried over (Mazenko and Valls 1983). It may be that the field of growth kinetics is just about to leave the stage of tentative probing.

Bibliography

Abbott, E.A. 1874. *Flatland*. Blackwell, Oxford (1974).

Abragam, A. and Goldman, M. 1982. *Nuclear Magnetism: Order and Disorder*. Clarendon, Oxford.

Abraham, F.F. 1982. Rep.Prog.Phys. **45**, 1113.

Adler, J. 1983. J.Phys.A **16**, 3585.

Adler, J. and Oitmaa, J. 1979. J.Phys.C **12**, 575.

Aharony, A. 1973a. Phys.Rev.B **8**, 4270.

Aharony, A. 1973b. Phys.Rev.B **8**, 4314.

Aharony, A. 1974. Phys.Rev.B **9**, 2416.

Aharony, A. and Bruce, A.D. 1979. Phys.Rev.Lett. **42**, 462.

Aharony, A. and Fisher, M.E. 1983. Phys.Rev.B **27**, 4394.

Aharony, A. and Halperin, B.I. 1975. Phys.Rev.Lett. **35**, 1308.

Alben, M.R. 1974. C.R.Acad.Sci.Ser.B **279**, 111.

Albrecht, O., Gruler, H., and Sackmann, E. 1978. J.Phys.(Paris) **39**, 301.

Alcaraz, F.C., Jacobs, L., and Savit, R. 1982. Phys.Lett. **89A**, 49.

Alcaraz, F.C., Jacobs, L., and Savit, R. 1983. J.Phys.A **16**, 175.

Alessandrini, V.A., Cracknell, A.P., and Przystawa, J.A. 1976. Comm.Phys. **1**, 51.

Alexander, S. and Pincus, P. 1980. J.Phys.A **13**, 263.

Allen, S.M. and Cahn J.W. 1979. Acta Metall. **27**, 1085.

Als-Nielsen, J. 1976. In *Phase Transitions and Critical Phenomena*. (Domb, C. and Green, M.S., eds.) Vol.V. p.87. Academic Press, London.

Ambler, E., Eisenstein, J.C., and Schooley, J.F. 1962. J.Math.Phys. **3**, 118.

Bak, P. 1982. Rep.Prog.Phys. **45**, 587.

Bak, P. 1983. Physics Today **36**. No.12, 25.

Bak, P. and Bohr, T. 1983. Phys.Rev.B **27**, 591.

Bak, P., Krinsky, S., and Mukamel, D. 1976a. Phys.Rev.Lett. **36**, 52.

Bak, P., Krinsky, S., and Mukamel, D. 1976b. Phys.Rev.Lett. **36**, 829.

Bak, P. and Mukamel, D. 1976. Phys.Rev.B **13**, 5086.

Bak, P. and Rasmussen F.B. 1981. Phys.Rev.B **23**, 4538.

Baker, G.A. and Hunter, D.L. 1973. Phys.Rev.B **7**, 3377.

Balzarini, D. and Mouritsen, O.G. 1983. Phys.Rev.A **28**, 3515.

Banavar, J.R., Grest, G.S., and Jasnow, D. 1980. Phys.Rev.Lett. **45**, 1424.

Banavar, J.R., Jasnow, D., and Landau, D.P. 1979. Phys.Rev.B **20**, 3820.

Baxter, R.J. 1971. Phys.Rev.Lett. **26**, 832.

Baxter, R.J. and Wu, F.Y. 1973. Phys.Rev.Lett. **31**, 1294.

Beauvillain, P., Chappert, C., and Laursen, I. 1980. J.Phys.A **13**, 1481.

Bell, G.M., Combs, L.L., and Dunne, L.J. 1981. Chem.Rev. **81**, 15.

Bernasconi, J. and Schneider, T. (eds.) 1981. *Physics in One Dimension*. Springer-Verlag, Heidelberg.

Berker, A.N. and Kadanoff, L.P. 1980. J.Phys.A **13**, L259.

Bervillier, C. 1976. Phys.Rev.B **14**, 4964.

Bhanot, G. and Creutz, M. 1980. Phys.Rev.B 22, 3370.

Binder, K. 1972. Physica 62, 508.

Binder, K. 1976. In *Phase Transitions and Critical Phenomena*. (Domb, C. and Green, M.S., eds.) Vol.Vb. p.1. Academic Press, New York.

Binder, K. (ed.) 1979. *Monte Carlo Methods in Statistical Physics I*. Springer-Verlag, Heidelberg.

Binder, K. 1981a. Z.Phys.B 45, 61.

Binder, K. 1981b. J.Stat.Phys. 24, 69.

Binder, K. 1983. Phys.Rev.Lett. 45, 811.

Binder, K. (ed.) 1984. *Monte Carlo Methods in Statistical Physics II*. Springer-Verlag, Heidelberg.

Binder, K. and Hohenberg, P.C. 1974. Phys.Rev.B 9, 2194.

Binder, K. and Kalos, M.H. 1979. In *Monte Carlo Methods in Statistical Physics I*. (Binder, K., ed.) p.225. Springer-Verlag, Heidelberg.

Binder, K. and Landau, D.P. 1980. Phys.Rev.B 21, 1941.

Binder, K. and Stauffer, D. 1974. Phys.Rev.Lett. 33, 1006.

Blankschtein, D. and Mukamel, D. 1982. Phys.Rev.B 25, 6939.

Blaschko, O., Ernst, G., Frantzl, P., Bernole, M., and Auger, P. 1982. Acta.Metall. 30, 547.

Bloch, D., Hermann-Ronzaud, D., Vettier, C., Yelon, W.B., and Alben, R. 1975. Phys.Rev.Lett. 35, 963.

Bloch, D., Vettier, C., and Burlet, P. 1980. Phys.Lett. 75A, 301.

Blöte, H.W.J. and Swendsen, R.H. 1980. Phys.Rev.B 22, 4481.

Blume, M., Heller, P., and Lurie, N.A. 1975. Phys.Rev.B 11, 4483.

Borsa, F., Pini, M.G., Rettori, A., and Tognetti, V. 1983. Phys.Rev.B 28, 5173.

Bortz, A.B., Kalos, M.H., and Lebowitz, J.L. 1981. J.Comp.Phys. 17, 10.

Brazovsky, S.A. and Dzyaloshinsky, I.E. 1975. Pis'ma Zh.Eksp.Teor.Fiz. 21, 360. [JETP Lett. 21, 164].

Brazovsky, S.A., Dzyaloshinsky, I.E., and Kukharenko, B.G. 1976. Zh.Eksp.Teor.Fiz. 70, 2257. [Sov.Phys.JETP 43, 1178].

Brézin, E., Le Guillou, J.C., and Zinn-Justin, J. 1973. Phys.Rev.D 8, 2418.

Brézin, E., Le Guillou, J.C., and Zinn-Justin, J. 1974. Phys.Rev.B 10, 892.

Brézin, E., Le Guillou, J.C., and Zinn-Justin, J. 1976. In *Phase Transitions and Critical Phenomena*. (Domb, C. and Green, M.S., eds.) Vol.VI. p.125. Academic Press, New York.

Brézin, E. and Zinn-Justin, J. 1976. Phys.Rev.B 13, 251.

Brinkmann, J., Courths, R., and Guggenheim, H.J. 1978. Phys.Rev.Lett. 40, 1286.

Bruce, A.D. and Wallace, D.J. 1976. J.Phys.A 9, 1117.

Bruinsma, R. and Aeppli, G. 1982. Phys.Rev.Lett. 48, 1625.

Buckingham, A.D., Disch, R.L., and Dunmur, D.A. 1968. J.Am.Chem.Soc. 90, 3104.

Burke, J.E., and Turnbull, D. 1952. In *Progress in Metal Physics*. (Chalmers, B., ed.) Pergamon, London.

Buzarè, J.Y., Fayet, J.C., Berlinger, W., and Müller, K.A. 1979. Phys.Rev.Lett. 42, 465.

Cadenhead, D.A., Müller-Landau, F., and Kellner, B.M.J. 1980. In *Ordering in Two Dimensions*. (Sinha, S.K., ed.) p.73. North Holland, New York.

Caillé, A., Pink, D.A., de Verteuil, F., and Zuckermann, M.J. 1980. Can.J.Phys. 58, 581.

Caillé, A., Rapini, A., Zuckermann, M.J., Cros, A., and Doniach, S. 1978. Can.J.Phys. 56, 348.

Callen, H.B. 1963. Phys.Lett. 4, 161.

Carpenter, H.C.H. and Elam, C.F. 1920. J.Inst.Metals 24, 83.

Chapman, D., Gomez-Fernandez, J.C., and Goni, F.M. 1979. FEBS Lett. **98**, 211.

Chapman, D., Williams, R.M., and Ladbrooke, B.D. 1967. Chem.Phys.Lipids **1**, 445.

Chappellier, M., Goldman, M., Vu Hoang Chau, and Abragam, A. 1969. C.R.Acad.Sci.Ser. B **268**, 1530.

Chen, S.C., Sturtevant, J.M. and Gaffney, B.J. 1980. Proc.Natl.Acad.Sci.USA **77**, 5060.

Ching, W.Y., Huber, D.L., Lagally, M.G., and Wang, G.-C. 1978. Surf.Sci. **77**, 550.

Chui, S.T., Forgacs, G., and Hatch, D.M. 1982. Phys.Rev.B **25**, 6952.

Chui, S.T. and Ma, K.B. 1983. Phys.Rev.B **27**, 4515.

Chung, T.T. and Dash, J.G. 1977. Surf.Sci. **66**, 559.

Cotterill, R.M.J. 1976. Biochim.Biophys.Acta **433**, 264.

Cox, S.F.J., Bouffard, V., and Goldman, M. 1975. J.Phys.C **8**, 3664.

Creutz, M. 1983. Phys.Rev.Lett. **50**, 1411.

Creutz, M., Jacobs, L., and Rebbi, C. 1979. Phys.Rev.D **20**, 1915.

Currie, J.F., Krumhansl, J.A., Bishop, A.R., and Trullinger, S.E. 1980. Phys.Rev.B **22**, 477.

Danielian, A. 1961. Phys.Rev.Lett. **6**, 670.

Danielian, A. 1964. Phys.Rev. **133**, A1344.

Davis, J.H., Bloom, M., Butler, K.W., and Smith, I.C.P. 1980. Biochim.Biophys.Acta **597**, 477.

Day, J. and Willis, C.R. 1981. J.Theor.Biol. **88**, 693.

Derrida, B., Pomeau, Y., Toulouse, G., and Vannimenus, J. 1980. J.Phys.(Paris) **41**, 213.

Diehl,R.D. and Fain, S.C. 1983. Surf.Sci. **125**, 116.

Diehl, R.D., Toney, M.F., and Fain, S.C. 1982. Phys.Rev.Lett. **48**, 177.

Ditzian, R.V., Banavar, J.R., Grest, G.S., and Kadanoff, L.P. 1980. Phys.Rev.B **22**, 2542.

Domany, E., Mukamel, D., and Fisher, M.E. 1977. Phys.Rev.B **15**, 5432.

Domany, E. and Riedel, E.K. 1978. Phys.Rev.Lett **40**, 561.

Domany, E. and Riedel, E.K. 1979. Phys.Rev.B **19**, 5817.

Domany, E. and Schick, M. 1979. Phys.Rev.B **20**, 3828.

Domany, E., Schick, M., Walker, J.S., and Griffiths, R.B. 1978. Phys.Rev.B **18**, 2209.

Domb, C. 1974. In *Phase Transitions and Critical Phenomena.* (Domb, C. and Green, M.S., eds.) Vol.III. p.357. Academic Press, New York.

Domb, C. and Green, M.S. (eds.) 1974. *Phase Transitions and Critical Phenomena.* Vol.III. Academic Press, New York.

Domb, C. and Green, M.S. (eds.) 1976. *Phase Transitions and Critical Phenomena.* Vol.VI. Academic Press, New York.

Doniach, S. 1978. J.Chem.Phys. **68**, 4912.

Doniach, S. 1980. In *Ordering in Two Dimensions.* (Sinha, S.K., ed.) p.67. North-Holland, North York.

Doucet, J. 1979. J.Phys.(Paris) Lett. **40**, L-185.

Eckert, J., Ellenson, W.D., Hastings, J.B., and Passell, L. 1979. Phys.Rev.Lett. **43**, 1329.

Elamrani, K. and Blume, A. 1983. Biochemistry **22**, 3305.

Erpenbeck, J.J. and Wood, W.W. 1977. In *Statistical Mechanics.* (Berne, B.J., ed.) Vol.6b. p.1. Plenum Press, New York.

Essam, J.W. and Sykes, M.F. 1963. Physica **29**, 378.

Feller, W. 1950. *An Introduction to Probability Theory and its Applications.* Vol. I. p.328. Wiley, New York.

Ferdinand, A.E. and Fisher, M.E. 1969. Phys.Rev. **185**, 832.

Ferer, M. and Velgakis, M.J. 1983. Phys.Rev.B **27**, 2839.

Fisher, M.E., 1960. Rep.Prog.Phys. **30**, 615.

Fisher, M.E. 1964. Am.J.Phys. **32**, 343.

Fisher, M.E. 1971. In *Proc. Int. School of Physics »Enrico Fermi«.* (Green, M.S., ed.) Course LI p.1. Academic Press, London.

Fisher, M.E. 1974. Rev.Mod.Phys. **46**, 597.

Fisher, M.E. 1981. Physica **106A**, 28.

Fisher, M.E. and Berker, A.N. 1982. Phys.Rev.B **26**, 2507.

Fisher, M.E. and Burford, R.J. 1967. Phys.Rev. **156**, 583.

Fisher, M.E. and Tarko, H.B. 1975. Phys.Rev.B **11**, 1217.

Fogedby, H.C., Hedegaard, P., and Svane, A. 1983. Phys.Rev.B **28**, 2893.

Forgacs, G. 1980. Phys.Rev.B **22**, 4473.

Fosdick, L.D. 1963. Meth.Comp.Phys. **1**, 245.

Frank, B. and Cheung, C.Y. 1984. Can.J.Phys. **62**, 35.

Frank, B., Cheung, C.Y., and Mouritsen, O.G. 1982. J.Phys.C **15**, 1233.

Frank, B. and Mitran, O. 1977. J.Phys.C **10**, 2641.

Frank, B. and Mouritsen, O.G. 1983. J.Phys.C **16**, 2481.

Frazer, B.C., Shirane, G., Cox, D.E., and Olsen, C.E. 1965. Phys.Rev. **140**, A1448.

Friedberg, R. and Cameron, J.E. 1970. J.Chem.Phys. **52**, 6049.

Frischleder, H. 1981. In *School Proceedings of the Sixth School on Biophysics of Membrane Transport.* (University of Warsaw, Warsaw) p.56.

Frowein, R. and Kötzler, J. 1976. Z.Phys.B **25**, 279.

Furukawa, H. 1984. Phys.Rev.A **29**, (July).

Fuselier, C.R., Gillis, N.S., and Raich, J.C. 1978. Solid St.Commun. **25**, 747.

Gaunt, D.S. and Guttmann, A.J. 1974. In *Phase Transitions and Critical Phenomena.* (Domb, C. and Green, D.S., eds.) Vol.III. p.181. Academic Press, New York.

Gaunt, D.S. and Sykes, M.F. 1973. J.Phys.A **6**, 1517.

Gaunt, D.S., Sykes, M.F., and McKenzie, S. 1979. J.Phys.A **12**, 871.

Georgallas, A. and Pink, D.A. 1982. J.Colloid Interface Sci. **89**, 107.

Gerling, R.W. and Landau, D.P. 1982. J.Appl.Phys. **53**, 7999.

Gitterman, M. and Mikulinsky, M. 1977. J.Phys.C **10**, 4073.

Glauber, R.J. 1963. J.Math.Phys. **4**, 294.

Glosli, J. and Plischke, M. 1983. Can.J.Phys. **61**, 1515.

Goldman, M. 1977. Phys.Rep. **32**, 1.

Goldman, M., Chapellier, M., Vu Hoang Chau, and Abragam, A. 1974. Phys.Rev.B **10**, 226.

Gordon, P. and El-Basyoumi, T.A. 1965. Trans.Metall.Soc. AIME **233**, 391.

Grant, M. and Gunton, D. 1983. Phys.Rev.B **28**, 5496.

Grest, G.S. and Banavar, J.R. 1981. Phys.Rev.Lett. **46**, 1458.

Grest, G.S. and Widom, M. 1981. Phys.Rev.B **24**, 6508.

Griffin, J.A., Huster, M., and Folweiler, R.J. 1980. Phys.Rev.B **22**, 4370.

Griffiths, H.P. and Wood, D.W. 1973. J.Phys.C **6**, 2533.

Griffiths, H.P. and Wood, D.W. 1974. J.Phys.C **7**, 4021.

Griffiths, R.B. 1972. In *Phase Transitions and Critical Phenomena.* (Domb, C. and Green, M.S., eds.) Vol.I. p.7. Academic Press, London.

Griffiths, R.B. 1981. Physica **106A**, 59.

Gunton, J.D., San Miguel, M., and Sahni, P.S. 1984. In *Phase Transitions and Critical Phenomena.* (Domb, C. and Lebowitz, J.L., eds.) Vol.VIII. p.267. Academic Press, London.

Güttinger, H. and Cannell, D.S. 1981. Phys.Rev.A **24**, 3188.

Haessner, F. (ed.) 1978. *Recrystallization of Metallic Materials.* Dr. Riederer Verlag, Stuttgart.

Halperin, B.I. and Hohenberg, P.C. 1977. Rev.Mod.Phys. **49**, 435.

Halperin, B.I. and Nelson, D.R. 1978. Phys.Rev.Lett. **41**, 121.

Hammersley, J.M. and Handscomb, D.C. 1967. *Monte Carlo Methods.* Methuen, London.

Handscomb, D.C. 1963. Proc.Camb.Phil.Soc. **60**, 115.

Harris, A.B. and Berlinsky, A.J. 1979. Can.J.Phys. **57**, 1852.

Harris, A.B., Mouritsen, O.G., and Berlinsky, A.J. 1984. Can.J.Phys.

Hastings, J.M. and Corliss, L.M. 1976. Phys.Rev.B **14**, 1995.

Hastings, J.M., Corliss, L.M., Kunnmann, W., Thomas, R., Begum, R.J., and Bak, P. 1980. Phys.Rev.B **22**, 1327.

Henderson, R. and Unwin, P.N.T. 1975. Nature **257**, 28.

Hennion, M., Ronzaud, D., and Guyot, P. 1982. Acta Metall. **30**, 599.

Hirsch, J.E. and Scalapino, D.J. 1983. Physics Today **36**. No.5. 44.

Hooghland, A., Spaa, J., Selman, B., and Campaner, A. 1983. J.Comp.Phys. **51**, 250.

Horwitz, G. and Callen, H. 1961. Phys.Rev. **124**, 1757.

Hullinger, F., Natterer, B., and Ott, H.R. 1978. J.Magn.Magn.Mater. **8**, 87.

Hunter, D.L. 1967. PhD Thesis, University of London (unpublished).

Hunter, D.L. and Baker, G.A. 1973. Phys.Rev.B **7**, 3346.

Hunter, D.L. and Baker, G.A. 1979. Phys.Rev.B **19**, 3808.

Imbro, D. and Hemmer, P.C. 1976. Phys.Lett. **57A**, 297.

Ising, E. 1925. Z.Phys. **31**, 253.

Israelachvili, J.N., Marcelja, S., and Horn, R.G. 1980. Quart.Rev.Biophys. **13**, 121.

Jacobs, L. and Rebbi, C. 1981. J.Comp.Phys. **41**, 203.

Jan, N., Lookman, T., and Pink, D.A. 1984. Biochemistry

Jayaprakash, C. and Tobochnik, J. 1982. Phys.Rev.B **25**, 4890.

José, J.V., Kadanoff, L.P., Kirkpatrick, S., and Nelson, D.R. 1977. Phys.Rev.B **16**, 1217.

Jüngling, K. 1976. Z.Phys.B **24**, 391.

Kadanoff, L.P. and Wegner, F.J. 1971. Phys.Rev.B **4**, 3989.

Kalia, R.K., Vashishta, P., and Mahanti, S.D. 1982. Phys.Rev.Lett. **49**, 676.

Kaski, K., Binder, K., and Gunton, J.D. 1983a. J.Phys.A **16**, L623.

Kaski, K. and Gunton, J.D. 1983. Phys.Rev.B **28**, 5371.

Kaski, K., Kumar, S., Gunton, J.D., and Rikvold, P.A. 1984. Phys.Rev.B

Kaski, K., Yalabik, M.C., Gunton, J.D., and Sahni, P.S. 1983b. Phys.Rev.B **28**, 5263.

Kawabata, C. and Binder, K. 1977. Solid St.Commun. **22**, 705.

Kawasaki, K. 1972. In *Phase Transitions and Critical Phenomena.* (Domb, C. and Green, M.S., eds.) Vol.II. p.443. Academic Press, London.

Kerszberg, M. and Mukamel, D. 1979. Phys.Rev.Lett. **43**, 293.

Kerszberg, M. and Mukamel, D. 1981. Phys.Rev.B **23**, 3943, 3953.

Ketley, I.J. and Wallace, D.J. 1973. J.Phys.A **6**, 1667.

Kirkwood, J.G. 1938. J.Chem.Phys. **6**, 70.

Kjaer, K., Nielsen, M., Bohr, J., Lauter, H.J., and McTague, J.P. 1982. Phys.Rev.B **26**, 5168.

Kjems, J.K., Passell, L., Taub, H., Dash, J.G., and Novaco, A.D. 1976. Phys.Rev.B **13**, 1446.

Kjems, J.K. and Steiner, M. 1978. Phys.Rev.Lett. **41**, 1137.

Knak Jensen, S.J. 1984. J.Phys.C

Knak Jensen, S.J. and Kjaersgaard Hansen, E. 1973. Phys.Rev.B **7**, 2910.

Knak Jensen, S.J. and Mouritsen, O.G. 1982. J.Phys.A **15**, 2631.

Knak Jensen, S.J., Mouritsen, O.G., Kjaersgaard Hansen, E., and Bak, P. 1979. Phys.Rev.B **19**, 5886.

Kogon, H.S. 1981. J.Phys.A **14**, 3253.

Kolb, M. 1983. Phys.Rev.Lett. **51**, 1696.

Kosterlitz, J.M., Nelson, D.R., and Fisher, M.E. 1976. Phys.Rev.B **13**, 412.

Kosterlitz, J.M. and Thouless, D.J. 1973. J.Phys.C **6**, 1181.

Kox, A.J., Michels, J.P.J., and Wiegel, F.W. 1980. Nature **287**, 317.

Kötzler, J. and Eiselt, G. 1976. Phys.Lett. **58A**, 69.

Krumhansl, J.A. and Schrieffer, J.R. 1975. Phys.Rev.B **11**, 3535.

Kubik, P.R. and Hardy, W.N. 1978. Phys.Rev.Lett. **41**, 257.

Kubik, P.R., Hardy, W.N., and Glattli, H. 1984. Surf.Sci.

Kumar, P. and Samalam, V.K. 1982. Phys.Rev.Lett. **49**, 1278.

Kushick, J. and Berne, B.J. 1977. In *Statistical Mechanics*. (Berne, B.J., ed.) Vol.6b. p.41. Plenum Press, New York.

Lagally, M.G., Wang, G.-C., and Lu, T.-M. 1978. CRC Crit.Rev.Solid St.Mater.Sci. **7**, 233.

Landau, D.P. 1976a. Phys.Rev.B **13**, 2997.

Landau, D.P. 1976b. Phys.Rev.B **14**, 255.

Landau, D.P. 1976c. Phys.Rev.B **14**, 4054.

Landau, D.P. 1977. Phys.Rev.B **16**, 4164.

Landau, D.P. 1979a. In *Monte Carlo Methods in Statistical Physics I*. (Binder, K., ed.) p.121. Springer-Verlag, Heidelberg.

Landau, D.P. 1979b. In *Monte Carlo Methods in Statistical Physics I*. (Binder, K., ed.) p.337. Springer-Verlag, Heidelberg.

Landau, D.P. 1980. Phys.Rev.B **21**, 1285.

Landau, D.P. and Binder, K. 1978. Phys.Rev.B **17**, 2328.

Landau, D.P. and Swendsen, R.H. 1981. Phys.Rev.Lett. **46**, 1437.

Landau, L.D. and Lifshitz, M.E. 1969. *Statistical Physics*. Pergamon, New York.

Larkin, A.I. and Khmel'nitskii, D.E. 1969. Zh.Eksp.Teor.Fiz. **56**, 2087 [JETP **29**, 1123].

Le Guillou, J.C. and Zinn-Justin, J. 1980. Phys.Rev.B **21**, 3976.

Lee, A.G. 1977. Biochim.Biophys.Acta **472**, 237.

Lenz, W. 1920. Phys.Z. **21**, 613.

Lethuillet, P., Pierre, J., Fillion, G., and Barbara, B. 1973. Phys.Status Solidi **15**, 613.

Leung, K.M. and Bishop, A.R. 1983. J.Phys.C **16**, 5893.

Levesque, D., Weiss, J.J., and Hansen, J.P. 1984. In *Monte Carlo Methods in Statistical Physics II*. (Binder, K., ed.) p.37. Springer-Verlag, Heidelberg.

Lieb, E.H., Schultz, T.D., and Mattis, D.C. 1961. Ann.Phys.(N.Y.) **16**, 407.

Liebman, R. 1981. Phys.Lett. **85A**, 59.

Liebman, R. 1982. Z.Phys.B **45**, 243.

Lifshitz, I.M. 1962. Zh.Eksp.Teor.Fiz. **42**, 1354 [Sov.Phys.JETP **15**, 939].

Lodish, H.F. and Rothman, J.E. 1979. Sci.Amer. **240**, 38.

Lookman, T. and Pink, D.A. 1984. Biochim.Biophys.Acta

Lookman, T., Pink, D.A., Grundke, E.W., Zuckermann, M.J., and de Verteuil, F. 1982. Biochemistry **21**, 5593.

Löser, W. and Sólyom, J. 1978. J.Phys.C **11**, 761.

Loveluck, J.M., Lovesey, S.W., and Aubry, S. 1975. J.Phys.C **8**, 3841.

Loveluck, J.M., Schneider, T., Stoll, E., and Jauslin, H.R. 1980. Phys.Rev.Lett. **45**, 1505.

Loveluck, J.M., Schneider, T., Stoll, E., and Jauslin, H.R. 1982. J.Phys.C **15**, 1721.

Ma, S.-k. 1976. Phys.Rev.Lett. **37**, 461.

Mabrey, S., Mateo, P.L, and Sturtevant, J.M. 1978. Biochemistry **17**, 264.

Magyari, E. and Thomas, H. 1983. J.Phys.C **16**, L535.

Maki, K. 1982. In *Progress in Low Temperature Physics*. (Brewer, D.F., ed.) Vol.III. p.1. North Holland, New York.

Marcelja, S. 1974. Biochim.Biophys.Acta **367**, 165.

Marro, J., Bortz, A.B., Kalos, M.H., and Lebowitz, J.L. 1975. Phys.Rev.B 12, 2000.

Martin, J.L. 1974. In *Phase Transitions and Critical Phenomena*. (Domb, C. and Green, M.S., eds.) Vol.III. p.97. Academic Press, New York.

Matveev, V.M. and Nagaev, E.L. 1972. Sov.Phys.Solid St. 14, 408.

Mazenko, G.F. and Valls, O.T. 1983. Phys.Rev.B 27, 6811.

McKenzie, S. and Gaunt, D.S. 1980. J.Phys.A 13, 1015.

McKenzie, S., Sykes, M.F., and Gaunt, D.S. 1979. J.Phys.A 12, 743.

Meirovitch, H. 1982. J.Phys.A 15, 2063.

Meyer, H.A. (ed.) 1956. *Symposium on Monte Carlo Methods*. Wiley, New York.

Metropolis, N., Rosenbluth, A., Rosenbluth, M., Teller, A., and Teller, E. 1953. J.Chem.Phys. 21, 1087.

Migone, A.D., Kim, H.K., Chan, M.H.W., Talbot, J., Tildesley, D.J., and Steele, W.A. 1983. Phys.Rev.Lett. 51, 192.

Mikeska, H.J. 1978. J.Phys.C 11, L29.

Mikeska, H.J. 1979. J.Magn.Magn.Mater. 13, 35.

Mikeska, H.J. and Osano, K. 1983. Z.Phys.B 52, 111.

Mouritsen, O.G. 1979. *High-temperature Series Analysis of Antiferromagnetic Phase Diagrams*. Ph.D. Thesis, Aarhus University (unpublished).

Mouritsen, O.G. 1980. J.Phys.C 13, 3909.

Mouritsen, O.G. 1983a. Biochim.Biophys.Acta 731, 217.

Mouritsen, O.G. 1983b. Phys.Rev.B 28, 3150.

Mouritsen, O.G. and Berlinsky, A.J. 1982. Phys.Rev.Lett. 48, 181.

Mouritsen, O.G. and Bloom, M. 1984. Biophys.J.

Mouritsen, O.G., Boothroyd, A., Harris, R., Jan, N., Lookman, T., MacDonald, L., Pink, D.A., and Zuckermann, M.J. 1983b. J.Chem.Phys. 79, 2027.

Mouritsen, O.G., Frank, B., and Mukamel, D. 1983a. Phys.Rev.B 27, 3018.

Mouritsen, O.G., Jensen, H., and Fogedby, H.C. 1984. Phys.Rev.B

Mouritsen, O.G., Kjaersgaard Hansen, E., and Knak Jensen, S.J. 1980. Phys.Rev.B 22, 3256.

Mouritsen, O.G. and Knak Jensen, S.J. 1978. Phys.Rev.B 18, 465.

Mouritsen, O.G. and Knak Jensen, S.J. 1979a. Phys.Rev.B 19, 3663.

Mouritsen, O.G. and Knak Jensen, S.J. 1979b. J.Phys.A 12, L339.

Mouritsen, O.G. and Knak Jensen, S.J. 1980a. Phys.Rev.B 22, 1127.

Mouritsen, O.G. and Knak Jensen, S.J. 1980b. Phys.Rev.B 22, 3271.

Mouritsen, O.G. and Knak Jensen, S.J. 1981. Phys.Rev.B 23, 1397.

Mouritsen, O.G., Knak Jensen, S.J., and Bak, P. 1977. Phys.Rev.Lett. 39, 629.

Mouritsen, O.G., Knak Jensen, S.J., and Bak, P. 1978. Ann.Israel Phys.Soc. 2, 548.

Mouritsen, O.G., Knak Jensen, S.J., and Frank, B. 1981a. Phys.Rev.B 23, 976.

Mouritsen, O.G., Knak Jensen, S.J., and Frank, B. 1981b. Phys.Rev.B 24, 347.

Mrozińska, A., Przystawa, J., and Sòlyom, J. 1979. Phys.Rev.B 19, 331.

Mukamel, D. 1976. Phys.Rev.Lett. 34, 481.

Mukamel, D. and Krinsky, S. 1976a. Phys.Rev.B. 13, 5065.

Mukamel, D. and Krinsky, S. 1976b. Phys.Rev.B 13, 5078.

Mukamel, D. and Wallace, D.J. 1979. J.Phys.C 13, L851.

Müller-Krumbhaar, H. 1979. In *Monte Carlo Methods in Statistical Physics I*. (Binder, K., ed.) p.195. Springer-Verlag, Heidelberg.

Nagle, J.F. 1975. J.Chem.Phys. 63, 1255.

Nagle, J.F. 1980. Ann.Rev.Phys.Chem. 31, 157.

Nath, K. and Frank, B. 1982. J.Appl.Phys. 53, 7971.

Nielsen, M., McTague, J.P., and Ellensen, W. 1977. J.Phys.(Paris) Colloq. L38, C4-10.

Nienhuis, B., Riedel, R.K., and Schick, M. 1983. Phys.Rev.B 27, 5625.

Nightingale, M.P. 1977. Phys. Lett. **59A**, 486.

Oitmaa, J. 1981. Can.J.Phys. **59**, 15.

Onsager, L. 1944. Phys.Rev. **65**, 117.

Opechowski, W. 1937. Physica **4**, 181.

O'Shea, S.F. and Klein, M.L. 1979. Chem.Phys.Lett. **66**, 381.

O'Shea, S.F. and Klein, M.L. 1982. Phys.Rev.B **25**, 5882.

Osheroff, D.D., Cross, M.C., and Fisher, D.S. 1980. Phys.Rev.Lett. **44**, 792.

Ott, H.R., Kjems, J.K., and Hullinger, F. 1979. Phys.Rev.Lett. **42**, 1378.

Palmer, R.G. 1982. Adv.Phys. **31**, 669.

Patkós, A. and Ruján, P. 1979. Z.Phys.B **33**, 163.

Pawley, G.S., Swendsen, R.H., Wallace, D.J., and Wilson, K.G. 1983. STATPHYS 15, 152.

Pearce, P.A. and Baxter, R.J. 1981. Phys.Rev.B **24**, 5295.

Pearce, C.J. 1978. Adv.Phys. **27**, 89.

Pearson, R.B., Richardson, J.L., and Toussaint, D. 1983. J.Comp.Phys. **51**, 241.

Penrose, O. 1978. In *Stochastic Processes in Nonequilibrium Systems.* (Garrido, L., ed.) p.210. Springer-Verlag, Heidelberg.

Pfeuty, P. and Toulouse, G. 1977. *Introduction to the Renormalization Group and Critical Phenomena.* Chap.9. Wiley, New York.

Phani, M.K., Lebowitz, J.L., and Kalos, M.H. 1980a. Phys.Rev.B **21**, 4027.

Phani, M.K., Lebowitz, J.L., Kalos, M.H., and Tsai, C.C. 1979. Phys.Rev.Lett. **42**, 577.

Phani, M.K., Lebowitz, J.L., Kalos, M.H., and Penrose, O. 1980b. Phys.Rev.Lett. **45**, 366.

Pink, D.A. 1983. In *Membrane Structure and Function.* (Chapman, D., ed.) McMillan Press, London.

Pink, D.A., and Carroll, C.E. 1978. Phys.Lett. **66A**, 157.

Pink, D.A. and Chapman, D. 1979. Proc.Natl.Acad.Sci.USA **76**, 1542.

Pink, D.A., Georgallas, A., and Zuckermann, M.J. 1980a. Z.Phys.B **40**, 103.

Pink, D.A., Green, T.J., and Chapman, D. 1980b. Biochemistry **20**, 6692.

Pink, D.A., Lookman, T., MacDonald, A.L., Zuckermann, M.J., and Jan, N. 1982. Biochim.Biophys.Acta **687**, 42.

Plischke, M. and Zobin, D. 1977. Can.J.Phys. **55**, 1126.

Quinn, P.J. and Chapman, D. 1980. CRC Crit.Rev.Biochem. **8**, 1.

Rácz, Z. and Vicek, T. 1983. Phys.Rev.B **27**, 2992.

Rahman, A. 1964. Phys.Rev. **136**, A405.

Ramirez, A.P. and Wolf, W.P. 1982. Phys.Rev.Lett. **49**, 227.

Ramsey, N.F. 1956. Phys.Rev. **103**, 20.

Ranck, J.L. 1983. Chem.Phys.Lipids **32**, 251.

Rasmussen, E.B. and Knak Jensen, S.J. 1981. Phys.Rev.B **24**, 2744.

Rebbi, C. 1980. Phys.Rep. **67**, 55.

Ritchie, D.S. and Fisher, M.E. 1972. Phys.Rev.B **5**, 2668.

Roelofs, L.D., Park, R.L., and Einstein, T.L. 1979. J.Vac.Sci.Technol. **16**, 478.

Roelofs, L.D., Kortan, A.R., Einstein, T.L., and Park, R.L. 1981. Phys.Rev.Lett. **46**, 1465.

Roger, M., Hetherington, J.H., and Delrieu, J.M. 1983. Rev.Mod.Phys. **55**, 1.

Rogiers, J., Betts, D.D., and Lookman, T. 1978. Can.J.Phys. **56**, 420.

Rogiers, J., Ferer, M., and Scaggs, E.R. 1979. Phys.Rev.B **19**, 1644.

Roinel, Y., Bachella, G.L., Avenel, O., Bouffard, V., Pinot, M., Roubeau, P., Meriel, P., and Goldman, M. 1980. J.Phys.(Paris) Lett. **41**, L-123.

Roinel, Y., Bouffard, V., Bachella, G.L., Pinot, M., Meriel, P., Roubeau, P., Avenel, O., Goldman, M., and Abragam, A. 1978. Phys.Rev.Lett. **41**, 1572.

Roskies, R.Z. 1981. Phys.Rev.B **24**, 5305.

Rudnick, J. 1978. Phys.Rev.B **18**, 1406.

Ruocco, M.J. and Shipley, G.G. 1982. Biochim.Biophys.Acta **691**, 309.

Rushbrooke, G.S. 1968. In *Physics of Simple Liquids.* (Temperley, H.N.V., Rowlingson, J.S., and Rushbrooke, G.S., eds.) p.25. North-Holland, Amsterdam.

Rushbrooke, G.S., Baker Jr., G.A., and Wood, P.J. 1974. In *Phase Transitions and Critical Phenomena.* (Domb, C. and Green, M.S., eds.) Vol.III. p.245. Academic Press, New York.

Sacco, J.E. and Chalupa, J. 1981. Solid St.Commun. **39**, 75.

Sadiq, A. 1984. J.Comp.Phys.

Sadiq, A. and Binder, K. 1983. Phys.Rev.Lett. **51**, 674.

Sadiq, A. and Binder, K. 1984. Z.Phys.B

Safran, S.A. 1981. Phys.Rev.Lett. **46**, 1581.

Safran, S.A., Sahni, P.S., and Grest, G.S. 1982. Phys.Rev.B **26**, 466.

Safran, S.A., Sahni, P.S., and Grest, G.S. 1983. Phys.Rev.B **28**, 2693.

Sahni, P.S., Dee, G., Gunton, J.D., Phani, M.K., Lebowitz, J.L., and Kalos, M.H., 1981. Phys.Rev.B **24**, 410.

Sahni, P.S., Grest, G.S., Anderson, M.P., and Srolovitz, D.J. 1983a. Phys.Rev.Lett. **50**, 263.

Sahni, P.S., Grest, G.S., and Safran, S.A. 1983c. Phys.Rev.Lett. **50**, 60.

Sahni, P.S. and Gunton, J.D. 1981. Phys.Rev.Lett. **47**, 1754.

Sahni, P.S., Srolovitz. D.J., and Grest, G.S. 1983b. Phys.Rev.B **28**, 2705.

Saito, Y. 1981. Phys.Rev.B **24**, 6652.

Sandermann, H. 1978. Biochim.Biophys.Acta **515**, 209.

Saul, D., Wortis, M., and Stauffer, D. 1974. Phys.Rev.B **9**, 4964.

Savit, R. 1980. Rev.Mod.Phys. **52**, 453.

Scalapino, D.J., Sears, M., and Ferrell, R.A. 1972. Phys.Rev.B **6**, 3409.

Schick, M. 1981. Phys.Rev.Lett. **47**, 1347.

Schick, M. 1982. Physica **109-110B**, 1811.

Schick, M. 1983. Surf.Sci. **125**, 94.

Schmidt, K.E. and Kalos, M.H. 1984. In *Monte Carlo Methods in Statistical Physics II.* (Binder, K., ed.) p.125. Springer-Verlag, Heidelberg.

Schneider, T. and Stoll, E. 1982. J.Appl.Phys. **53**, 8024.

Schulhof, M.F., Heller, P., Nathans, R., and Linz, A. 1970. Phys.Rev.B **1**, 2304.

Scott, H.L. 1977. Biochim.Biophys.Acta **469**, 264.

Scott, H.L. and Cherng, S.L. 1978. Biochim.Biophys.Acta **510**, 209.

Scott, T.A. 1976. Phys.Rep. **27**, 89.

Seelig, J. 1981. In *Membranes and Intercellular Communication.* (Balian, R. et al. , eds.) p.15. North-Holland, New York.

Seelig, J. and Seelig, A. 1980. Quart.Rev.Biophys. **13**, 19.

Selke, W., Binder, K., and Kinzel, W. 1983. Surf.Sci. **125**, 74.

Selke, W. and Fisher, M.E. 1979. Phys.Rev.B **20**, 257.

Shenker, S.H. and Tobochnik, J. 1980. Phys.Rev.B **22**, 4462.

Shugard, W.J., Weeks, J.D., and Gilmer, G.H. 1980. Phys.Rev.B **21**, 5309.

Sinha, S.K. (ed.) 1980. *Ordering in Two Dimensions.* North-Holland, New York.

Smart, J.S. 1966. *Effective Field Theories of Magnetism.* W.B. Saunders Co., London.

Snyder, B. and Freire, E. 1980. Proc.Natl.Acad.Sci.USA **77**, 4055.

Sólyom, J. and Grest, G.S. 1977. Phys.Rev.B **16**, 2235.

Sprenkels, J.C.M, Wenckebach, W.T., and Poulis, N.J. 1983. J.Phys.C **16**, 4425.

Srolovitz, D.J., Anderson, M.P., Grest, G.S., and Sahni, P.S. 1983. Scripta Metall. **17**, 241.

Stanley, H.E. 1971. *Introduction to Phase Transitions and Critical Phenomena.* Clarendon Press, Oxford.

Stauffer, D. 1982. J.Appl.Phys. **53**, 7980.

Steele, W.A. 1977. J.Phys.(Paris) Colloq. **38**, C4-61.

Steiner, M., Kakurai, K., and Kjems, J.K. 1983. Z.Phys.B **53**, 117.

Stoll, E., Binder, K., and Schneider, T. 1973. Phys.Rev.B **8**, 3266.

Sullivan, N.S. and Vaissiere, J.M. 1983. Phys.Rev.Lett. **51**, 658.

Suzuki, M. 1965. Phys.Lett. **19**, 267.

Suzuki, M. 1976. Prog.Theor.Phys. **56**, 1454.

Swendsen, R.H. 1979. Phys.Rev.Lett. **42**, 859.

Swendsen, R.H. 1982. In *Real Space Renormalization.* (Burkhardt, T. and van Leeuven, J.M.J., eds.) p.57. Springer-Verlag, Heidelberg.

Sykes, M.F., Gaunt, D.S., Roberts, P.D., and Wyles, J.A. 1972. J.Phys.A **5**, 640.

Syozi, I. 1972. In *Phase Transitions and Critical Phenomena.* (Domb, C. and Green, M.S., eds.) Vol.I. p.270. Academic Press, New York.

Taggert, G.B. and Fittipaldi, I.P. 1982. Phys.Rev.B **25**, 7026.

Tessier-Lavigne, M., Boothroyd, A., Zuckermann, M.J., and Pink, D.A. 1982. J.Chem.Phys. **76**, 4587.

Tobochnik, J. and Chester, G.V. 1979. Phys.Rev.B **20**, 3761.

Tognetti, V., Rettori, A., Pini, M.G., Loveluck, J.M., Balucani, V., and Balcar, A. 1983. J.Phys.C **16**, 5641.

Tsong, T.Y. and Kanehisa, M.I. 1977. Biochemistry **16**, 2674.

Tucciarone, A., Lau, H.Y., Corliss, L.M., Delapalme, A., and Hastings, J.M. 1971. Phys.Rev.B **4**, 3206.

Tyson, J.A. and Douglass Jr., D.H. 1966. Phys.Rev.Lett. **17**, 472.

Van der Ploeg, P. and Berendsen, H.J.C. 1982. J.Chem.Phys. **76**, 3271.

Velgakis, M.J. and Ferer, M. 1983. Phys.Rev.B **27**, 401.

Villain, J. 1959. J.Phys.Chem.Solids **11**, 303.

Villain, J. 1975. J.Phys.(Paris) **36**, 581.

Villain, J. 1977. J.Phys.C **10**, 1717.

Villain, J. and Gordon, M.B. 1980. J.Phys.C **13**, 3117.

Wang, G.-C., and Lu, T.-M. 1983. Phys.Rev.Lett. **50**, 2014.

Wannier, G.H. 1945. Rev.Mod.Phys. **17**, 50.

Wannier, G.H. 1950. Phys.Rev. **79**, 357.

Weinberg, W.H. 1983. Ann.Rev.Phys.Chem. **34**, 217.

Wegner, F.J. 1972. Phys.Rev.B **5**, 4529.

Wegner, F.J. and Riedel, F.J. 1973. Phys.Rev.B **7**, 248.

Widom, B. 1965. J.Chem.Phys. **43**, 3898.

Williams, E.D., Cunningham, S.L., and Weinberg, W.H. 1978. J.Chem.Phys. **68**, 4688.

Wilson, K.G. 1971. Phys.Rev.B **4**, 3174, 3184.

Wilson, K.G. 1983. Rev.Mod.Phys. **55**, 583.

Wilson, K.G. and Fisher, M.E. 1972. Phys.Rev.Lett. **28**, 240.

Wilson, K.G. and Kogut, J. 1974. Phys.Rep. **12C**, 75.

Wittington, S.G. and Chapman, D. 1966. Trans.Faraday Soc. **62**, 3319.

Wood, D.W. 1972. J.Phys.C **5**, L181.

Wood, D.W. and Griffiths, H.P. 1974. J.Phys.C. **7**, L54.

Wood, W.W. 1968. In *Physics of Simple Liquids.* (Temperley, H.N.W., Rowlingson, J.S., and Rushbrooke, G.S., eds.) p.115. North-Holland, Amsterdam.

Wortis, M. 1974. In *Phase Transitions and Critical Phenomena.* (Domb, C. and Green, M.S., eds.) Vol.III. p.113. Academic Press, New York.

Wu, F.Y. 1971. Phys.Rev.B **4**, 2312.

Wu, F.Y. 1972. Phys.Lett.A **38**, 77.

Yang, C.N. 1952. Phys.Rev. **85**, 809.

Yang, C.N. and Lee, T.D. 1952. Phys.Rev. **87**, 404.
Yang, C.-P. 1963. Proc.Symp.Appl.Math. **15**, 351.
Zhang, H.I. 1981. J.Phys.C **14**, 57.
Zia, R.K.P. and Wallace, D.J. 1975. J.Phys.A **8**, 1495.

Subject Index

Springer Series in

Information Sciences

Editors: K.-s. Fu, T. S. Huang, M. R. Schroeder

Springer-Verlag
Berlin
Heidelberg
New York
Tokyo